782520 1/74

14⁰⁰

an
introduction to
X-ray spectrometry

an
introduction to
X-ray spectrometry

RON JENKINS, 1932-

PHILIPS ELECTRONIC INSTRUMENTS
NEW YORK

London · New York · Rheine

Heyden & Son Ltd., Spectrum House, Alderton Crescent, London NW4 3XX.
Heyden & Son Inc., 225 Park Avenue, New York, N.Y. 10017, U.S.A.
Heyden & Son GmbH, 4440 Rheine/Westf., Münsterstrasse 22, Germany.

Library of Congress Catalog Card No. 73-93365
ISBN 0 85501 035 5

Printed in Great Britain by J. W. Arrowsmith Ltd., Bristol BS3 2NT

contents

CHAPTER 3   physics of X-rays

CHAPTER 4   instrumentation

CHAPTER 8 the study of chemical bonding

preface

It is now seven years since the publication of "Practical X-Ray Spectrometry" by Drs de Vries and myself and four years since the introduction of our second book on "Worked Examples in X-Ray Analysis". It had always been the intention to prepare a series of four books dealing with respectively, an introduction to the method, practical problems dealing mainly with instrumentation, a volume dealing with worked examples and data manipulation, and finally a volume devoted purely to quantitative methods. It was apparent to us, from our involvement in schools, conferences and colloquia, that the initial priorities should be the second and third of these volumes with the others following later. Our experience at that point in time had been that good qualitative and quantitative analysis was more likely to stem from a thorough appreciation of the limitations of the instrumentation and a familiarity with standard data handling methods, rather than a deep understanding of the theory of X-ray physics.

In the succeeding years the X-ray spectrometer has become well established as a standard tool in many analytical laboratories and we seem now to be well advanced into the period of consolidation. It is clear that a better understanding of the fundamentals of the technique would be of great benefit to many of the analysts employing X-ray methods and it is to be hoped that this new work will provide this information.

Particular attention has been given to the areas of X-ray spectra, the use of X-ray methods for the study of chemical bonding and the development of algorithms for the conversion of count data to concentration. It is hoped that the book will also be useful to beginners in the field since the sections on instrumentation, errors and qualitative analysis are treated in a fairly general way and as such should give a good understanding of the basic principles.

This book is thus not intended to replace previous works but rather to supplement them. Some overlap is inevitable, however, and advantage has been taken of this to give more basic information about energy dispersive methods which have come into great prominence over the past few years. Instrumentation and methodology for both crystal spectrometers and energy dispersive spectrometers are described, and an attempt has been made to compare and contrast the relative merits of the two systems.

I should like to acknowledge the help of friends and colleagues in the preparation of this work, in particular, Drs Hans de Vries and Dr. Frank Paolini, both of whom have demonstrated over the years that they are willing to argue a fine point to the bitter end. Others who helped with the reading of the manuscript include Dr. Russell Westberg, Dr. David Haas, John Croke and Irene Piscopo-Rodgers. Finally, special thanks are due to my secretary, Rose Gaw, whose great patience in typing the manuscript is gratefully acknowledged.

RON JENKINS

New York, March 1974

symbols and terminology

λ	wavelength	F	structure factor
E	energy	x	path length of radiation
θ	the (Bragg) diffraction angle	d	penetration depth
ψ_1	angle between specimen surface and centre of X-ray beam	C	concentration
		W	weight fraction
ψ_2	take-off angle of spectrometer (angle between axis of collimator and specimen surface)	ω	fluorescent yield
		I	counting rate
		p	peak position
k_{ij}	correction factor for element j on element i using the intensity for correction	b	background position
		N	number of counts
		A	$\sin \psi_1 / \sin \psi_2$
α_{ij}	correction factor for element j on element i using the concentration for correction	d_{hkl}	interplanar spacing
		T	time
		t	dead time
m	counts/s per percent	V	potential
Z	atomic number	V_0	critical excitation potential
σ	scattering coefficient	F	Fano factor
s	standard deviation	ϕ	binding energy
τ	photoelectric absorption	g	probability of electron transfer within a particular series
h	Planck's constant		
v	frequency	Γ	detector resolution
μ	mass absorption coefficient	$J(\lambda)$	X-ray tube spectrum
ρ	density	r	absorption jump
μ^*	linear absorption coefficient	i	current
f	atomic scattering factor	c	velocity of light

CHAPTER 1
introduction

1.1   THE PLACE OF THE SPECTROMETER IN THE MODERN LABORATORY

The present day analytical chemist has many methods available to him for the qualitative and quantitative determination of elemental composition. These methods include a large number of instrumental techniques which, although for the most part only comparative, generally offer great sensitivity, speed, ease of automation and high precision. Where suitable calibration procedures are employed, they may also offer great accuracy. Foremost among these instrumental techniques are the spectroscopic methods which are, in general terms, based on the correlation of elemental composition with specific intensity variations in the emission or absorption spectra of the excited specimen. Since these intensity variations have generally to be made at discrete wavelengths, the heart of the instrument is a spectrometer, this being a device which allows a certain wavelength, or wavelength range, to be selected for measurement. A "spectrometer", in the generally accepted sense of the word, consists not only of the spectrometer itself, but also a primary radiation source, a specimen handling device and a signal measuring and read-out system.

It is unlikely that a single type of spectrometer would provide a given laboratory with the full capability required, in terms of range of elements measurable, sensitivity, specimen, throughput, etc., and a well equipped facility will generally have several types of spectrometer available. However, the high cost of modern instrumentation often restricts the smaller laboratory to one or two spectrometer types and, for this reason, the most successful spectrometers have been those which offer great versatility in terms of the range of elements coverable, coupled with high precision and sensitivity. Two of the most successful emission spectrometers which fit into this category are the ultraviolet emission spectrometer and the X-ray fluorescence spectrometer. These spectrometers derive their names from the wavelength region in which they operate. Thus the u.v. spectrometer operates in the ultraviolet region where the characteristic emission wavelengths originate from outer atomic electrons, excited to partially or unfilled atomic or molecular orbitals, which revert back to their initial or

1

"ground" state. The X-ray spectrometer operates in the X-ray region where characteristic emission wavelengths originate from removal of inner orbital electrons, followed by de-excitation to the ground state by transference of outer shell electrons to the inner shell vacancies. In each case, the emission wavelengths are dependent upon the electronic configuration of the excited atom (that is, on the atomic number) and hence each atom will emit a characteristic emission spectrum. Since, in general, there are many more unfilled atomic and molecular orbitals than electrons in the inner shells, u.v. emission spectra are more complex than X-ray spectra. Also, X-ray spectra are for the most part independent of state of chemical combination of the excited element. Against this, however, every element in the periodic table will, in principle at least, give a measurable u.v. spectrum, whereas X-ray spectra from the very low atomic number elements are more dependent upon molecular orbitals than on atomic orbitals, and both the "sharpness" and unique character of the corresponding emission lines tend to deteriorate. Thus, in practical terms, the u.v. emission spectrometer allows the determination of most of the elements in the periodic table including such important low atomic number elements as carbon and boron. The X-ray spectrometer, although also covering the greater part of the atomic number range, is difficult to use for the measurement of elements below atomic number 9 (fluorine): hence carbon, boron, beryllium, etc., are precluded. On the other hand, inter-element interferences (matrix effects) are more predictable and more easily corrected for in the X-ray region, and wide concentration range calibration is easier to achieve.

This present text is devoted exclusively to X-ray spectrometry and for detailed discussion of other spectroscopic methods, the reader is advised to consult some of the many publications dealing with other techniques.

1.2 X-RADIATION AND X-RAY SPECTROMETRY

The X-ray region is that part of the electromagnetic spectrum between about 0.1 and 200 Å (Ångstrom units) where $Å = 10^{-10}$ m. It is thus bounded by the (conventionally accepted) gamma-ray region to the short wavelength side, and by the vacuum ultraviolet region to the long wavelength side. Only a relatively small part of the total X-ray region is covered by the conventional X-ray spectrometer, about 0.2–20 Å, this being referred to as the analytical X-ray region. Wavelengths shorter than a few tenths of an Ångstrom are difficult to excite and even more difficult to disperse or separate. Wavelengths in excess of 20 Å arise mainly from outer orbital transitions and are less useful for element characterization.

The dispersing medium in an X-ray spectrometer differs from the prism used in the visible or ultraviolet region, and from the diffraction grating used in the vacuum ultraviolet. Most X-ray spectrometers are based either on the diffracting properties of certain single crystals, such as quartz and lithium fluoride, or on the proportional characteristics of certain photon counters such as the lithium drifted silicon detector. The former system is the basis of the so-called "wavelength dispersion" or "crystal dispersion" spectrometer and the latter is the basis of the "energy-dispersion" spectrometer.

All methods of analysis based on X-ray spectrometry involve the excitation of characteristic wavelengths from the elements making up the sample being analysed, separation of these wavelengths by means of a spectrometer, measurement of the intensities of the individual characteristic wavelengths and estimation of elemental composition by use of the measured intensities. X-Ray spectrometry is by no means a new technique and its potential use as a means of qualitatively identifying elements has been recognized since the early 1900's. Indeed, X-ray spectroscopic measurements were uniquely responsible for the identification of several previously undiscovered elements in the periodic classification. Reasonably good quantitative data were being obtained by the early 1920's even though the best measuring devices available at the time were photographic film and the ionization chamber. The present-day X-ray crystal spectrometer does not differ too much, at least in principle, from the early equipment, although most modern instruments utilize an X-ray source for the excitation of secondary X-radiation from the sample (hence the term "fluorescence"), rather than the direct or indirect electron excitation used earlier. Also, the film technique has been completely replaced by photon counters: hence one should refer to the instrument as a spectrometer rather than a spectrograph. The rapid growth of modern X-ray spectrometry has come mainly as a result of the improvements in detection systems brought about in the early 1940's, coupled with the increased efforts of scientists to provide rapidly growing industry with fast and reproducible methods of analysis.

1.3   EARLY DEVELOPMENT OF X-RAY SPECTROSCOPIC ANALYSIS

Early in 1896, Roentgen made available a report on his discovery of X-rays. This report "Ueber eine neue Art von Strahlen" described the initial discovery of the radiation and subsequent work on its absorption properties.[1] From this work has grown four major fields of applications which, over the past fifty years, have affected the course of medicine, industry and research. The first two of these areas are medical and industrial radiography, of which the medical application was apparent from the very birth of X-ray technology. Indeed the first report from Roentgen contained, as experimental evidence, a radiograph of his wife's hand. The third of the techniques is X-ray diffraction which began in 1912 with the classic experiment of Friedrich and Knipping, under the direction of Max von Laue,[2] in which the diffraction property of crystalline materials and the wave character of X-rays were confirmed for the first time. The fourth technique is X-ray spectrometry. As has already been mentioned, X-ray spectrometry is an old technique, dating back to the early

1900's, but one which has developed as an accepted potent analytical technique only over the past fifteen years or so.

The first measurements which indicated that X-ray emission was associated with atomic number were performed by Barkla,[3] who demonstrated that different "hardnesses" (a term used to denote penetrating power) of radiation were emitted, depending upon excitation and target characteristics. It was in fact Barkla who proposed the line series nomenclature K, L, M, etc., still in use today. But the worker to whom much credit must go for the confirmation of the relationship between wavelength and atomic number is H. G. J. Moseley. The work of Moseley in 1913 must be considered as a bringing together of many half-formed ideas in the minds of several of his contemporary scientists. At that time, Niels Bohr had just extended Rutherford's concept of atomic structure, suggesting that a central nucleus surrounded by discrete electron "shells" should be considered. Bohr had recently visited Rutherford[4] at Manchester and it was Rutherford who was to discuss the problem with one of his young demonstrators (Moseley). The Braggs had just demonstrated the "reflection" method of diffraction with their new spectrometer (which was really akin to what we would now call a diffractometer) and it was this technique that Moseley was to employ.

Further, they offered an explanation of the diffraction condition which was far less cumbersome than that offered by von Laue, whose treatment involved a mathematical analysis of the three dimensional lattice. Moseley recorded the emission spectra from a range of elements and postulated the "Moseley Law",[5] which relates the emission wavelength (λ) with atomic number (Z) in the form

$$\frac{1}{\lambda} = K[Z - \sigma]^2 \tag{1-1}$$

in which K is a constant which depends upon the spectral series, and σ is the shielding constant. Moseley was tragically killed in the Dardanelles campaign of 1915, but his mark on X-ray spectrometry is immortal.[6]

In the 1920's and 1930's X-ray spectroscopy played an enormous part in the clarification of the periodic classification of elements and X-ray measurements were directly responsible for the discovery of the elements hafnium and rhenium. The analytical possibilities of the technique were also recognized at a very early stage and the first recorded analytical application of the X-ray method is the work of Hadding, who in 1922 described his method for the analysis of minerals.[7] ·· ·re are many fascinating accounts of the early successes of analytical X-ray spectrometry[8–11] and the great names in X-ray spectroscopy in the pre-war era include Siegbahn, von Hevesey, Blokkin, Laby and many others.

The instrumentation available at that time was mostly home-made although at least one spectrometer became commercially available in the late 1930's.[12] A typical spectrometer generally used a direct or scattered electron beam source, an analysing crystal—usually rock salt or calcite—and a film as the measuring device. The spectrometer was evacuable and quite capable of measurements in the 0.3 to 50 Å region. Sensitivities of a few tens of parts per million could be attained in favourable cases. The biggest experimental problem was probably that of specimen presentation, because for the most part the specimen was bathed by the electron beam from the source, and was thus subject to heating and high vacuum, with subsequent problems of volatility and melting. The advantages of fluorescence excitation were recognized in the early 1920's but this was generally employed in the open window type configuration described on p. 53.

The first suggestion to use primary X-rays, rather than electrons, as an excitation source, appears to have come from von Hevesey who, in 1925, suggested this approach to Coster and Nishina.[13] It is often stated that the credit for the initial "discovery" of the fluorescence technique should go to Glocker and Schreiber, but it appears that this work was not reported until 1928.[14] In any event, the sensitivities available at that time did not encourage the use of an excitation technique which, although more convenient from the point of view of specimen presentation, was less efficient by several orders of magnitude. It is, however, interesting to quote from von Hevesey's book "Chemical Analysis by X-rays and its Applications" published in 1932:

"... The great advantages of the secondary method are entire avoidance of the disturbing cathode ray effects, the possibility of investigating even liquids, and the almost complete absence of the continuous background on the photographic plate. The latter is a very great advantage, for when using the primary method with weak lines it is necessary to limit the exposure, otherwise such lines become entirely covered by the continuous background, and a longer exposure does not help. Having practically no continuous background from the secondary X-ray method, we can expose a plate even at an exceedingly long time, the difference between the intensity of the line and of the background constantly increasing. ..."

"... Secondary analysis would be much facilitated by the manufacture of sealed X-ray tubes fitted with a window of beryllium or the like to permit the passage of the soft rays often desired in X-ray analysis. An important requirement would be to have an entire cathode and window nearer together .... "

The ideas of von Hevesey and others were to lie dormant until the late 1940's, when again, as in 1913, new ideas and improved technology, plus the experimental ingenuity of Friedman and Birks,[15] came together to produce the embryo of the "modern" X-ray spectrometer. Telecommunications development in World War Two required the provision of accurately orientated quartz oscillator plates, and it was the spectrometer designed to orientate these plates which was to serve as the basis of the Friedman–Birks spectrometer. The slow and inconvenient film measuring method was replaced by the thin window Geiger counter and its associated electronic counting gear. A sealed, beryllium window, X-ray tube was employed, this coming about as a result of fifty years of development in the radiographic field. A collimator system made of nickel tubes and a specimen presentation device completed the spectrometer, which is remarkably similar in almost every detail to its present day counterpart.

The significance of this development was quickly recognized and "Geiger-counter spectrometers" became commercially available within the next couple of years, first as an attachment to powder diffractometers, but later as self contained units. Vacuum spectrometers followed quickly afterwards, along with the gas flow proportional counter and the lithium fluoride crystal. In 1950 there were probably about one hundred X-ray spectrometers in use. By 1960 this was one thousand and in 1970 nearly ten thousand.

The advent of the semiconductor-based energy dispersion spectrometer in the mid to late 1960's has proven to be another breakthrough and several thousand of these systems are currently in use.

1.4   THE MODERN X-RAY SPECTROMETER

Modern X-ray spectrometers are of many types ranging from cheap, simple systems for the measurement of a single element to complex, expensive, auto-mated systems capable of providing thousands of determinations per week. All contain a source, a dispersion system and a detector. Several broad cate-gories exist, the two major divisions being sequential (scanning) and simultaneous (multi-channel) instruments. A sequential instrument scans over a selected wavelength range either continuously or incrementally between fixed wavelength positions. In the simultaneous spectrometer, as many spectrometers are provided as there are wavelengths to be measured and all wavelengths are measured at the same time. The scanning type is much slower than the multichannel instru-ment, but on the other hand is more flexible and is equally applicable to both quantitative and qualitative work. A given instrument may have both fixed and scanning channels where the increased cost is justified by the need for both speed and flexibility. In either case, a crystal or energy dispersion spectrometer may provide the basis of the instrumentation. Simple single-element spectrometers are generally based on a radioisotope source plus a pair of "balanced" filters as the means of wavelength isolation. The "band-pass" between the filters is selected to cover the required wavelength.

Automatic data handling is now becoming relatively common-place as is the provision of multiple sample handling facilities. In general terms it can be stated that the modern X-ray spectrometer is capable of providing data on all elements in the periodic table above oxygen ($Z = 8$), with detection limits in the low parts per million range. Instrumental precision is of the order of 0.1 % and analytical accuracies in the range 0.2–1 % relative are obtainable following correct calibration and data handling. A sequential spectrometer will provide analysis on prepared samples at the rate of one per 30–100 seconds, while some simultaneous units can analyse for up to 30 elements in 1–2 minutes. The technique is in principle non-destructive, although some preparation of the sample is invariably necessary. The cost of instrumentation ranges from about $10,000 to $150,000.

X-Ray spectrometry is considered by many to stand head and shoulders above all other instrumental techniques of elemental analysis. Its speed, accuracy, sensitivity and flexibility, coupled with the capability to cover the greater part of the periodic classification, have made it an invaluable tool in nearly all branches of science. At the present time, between three and four hundred X-ray spectrometry publications a year appear in the scientific press—a sure indication of its acceptance in the field of materials analysis.

References

(1) Roentgen, W. C., *Ann. Phys. Chem.* **64**, 1 (1898).
(2) Friedrich, W., Knipping, P. and von Laue, M., *Ber. Bayer. Akad. Wiss.* 303 (1912); *Ann. Physik* **41**, 971 (1913).
(3) Barkla, C. G., *Phil. Mag.* **22**, 396 (1911).
(4) Oliphant, M., *Rutherford, Recollections of the Cambridge Days*, Elsevier, Amsterdam (1972).
(5) Moseley, H. G. J., *Phil. Mag.* **26**, 1024 (1912); **27**, 703 (1913).
(6) Jaffe, B., *Moseley and the Numbering of the Elements*, Doubleday, New York (1971).
(7) Hadding, A., *Z. Anorg. Allgem. Chem.* **122**, 195 (1922).
(8) Von Hevesey, G., *Chemical Analysis by X-rays and its Applications*, McGraw-Hill, New York, 1932.
(9) Siegbahn, Manne, *The Spectroscopy of X-rays*, Oxford University Press, Oxford, 1925.
(10) Blokkin, M. A., *Methods of X-ray Spectroscopic Research*, Pergamon, London, 1965.
(11) Siegbahn, Manne, in Edward, P. P. (Ed.), *Fifty Years of X-ray Diffraction*, Oosthoek, Utrecht, 1962, Chapter 16.
(12) *Professor Laby's X-ray Spectrograph*, Hilger Publication No. S.B. 106/3 (1938).
(13) Coster, D. and Nishina, J., *Chem. News*, **130**, 149 (1925).
(14) Glocker, R. and Schrieber, H., *Ann. Physik* **85**, 1085 (1928).
(15) Friedman, H. and Birks, L. S., *Rev. Sci. Instr.* **19**, 323 (1948).

CHAPTER 2

X-ray spectra

When an element is bombarded with primary photons, electrons from the atomic subshells may be excited to unfilled orbital levels. The element regains its initial or "ground" state by transference of outer orbital electrons to the unfilled inner levels and the energy surplus following each transference may be emitted as characteristic radiation. The energy of the emitted X-ray photon is equal to the absolute energy difference between the binding energies of the initial and final states of the transferred electron. Thus the wavelength of the photon is dependent upon the distribution of electrons in the excited atom and hence upon the atomic number of that atom. Since, under a given set of circumstances, many such electron transferences take place simultaneously, several characteristic lines are emitted at the same time and we refer to this total emission as the characteristic emission spectrum of the element. A fairly simple set of selection rules covers the normal transitions (diagram lines) but certain lines are observed which do not apparently fit the selection rules. These lines are categorized as forbidden and satellite lines and, although they are generally weak, they do have significance in analytical X-ray spectroscopy.

The following sections discuss the electronic configurations of the elements and the various selection rules for the formulation of transitions and diagram lines. Characteristics of diagram lines, satellites and forbidden transitions are discussed and illustrated with examples taken from the various wavelength series.

2.1 BASIC UNITS

In X-ray spectrometry, it is necessary to utilize both wavelength and energy terminology to explain the properties of X-radiation. The most common units utilized are the Ångstrom unit for wavelength and the kilo electron volt for energy.

The Ångstrom unit (Å) is 10^{-10} of a metre (the metre was accurately redefined in 1960 as 1,650,763.73 wavelengths *in vacuo* of a particular emission line of krypton-86). X-Ray spectroscopists have, however, always used their own wavelength standards, and for nearly forty years most X-ray wavelengths were expressed in terms of the "X unit", first introduced by Siegbahn.[1] The X unit is based on the first order grating constant of calcite (d_1 = 3029.04 XU) and was intended to be 10^{-3} Å. Unfortunately, not all calcite crystals are identical and errors of about 5–10 ppm can be introduced, thus many earlier wavelength measurements were not self-consistent. Since 1960, Bearden and his co-workers at Johns Hopkins University in Baltimore have sought to rationalize and re-measure X-ray wavelengths[2] relative to a selected primary standard.[3] The primary wavelength standard taken by Bearden is the $K\alpha_1$ line of tungsten and this is supplemented by four secondary wavelength standards, namely Ag $K\alpha_1$, Mo $K\alpha_1$, Cu $K\alpha_1$ and Cr $K\alpha_2$. Table 2-1 lists the actual values of the primary and four secondary standards. One of the greatest advantages to be gained by use of W $K\alpha_1$ as a primary standard, is the ability to measure it in transmission; thus refraction and anomalous dispersion effects are negligible. Since the accurate measurement of angles by interferometry is also best suited to the transmission method, this is a further benefit. Also, the W $K\alpha_1$ line is very symmetrical and chemical shifts due to isotopic effects are negligible.

The basic unit for energy is the kiloelectron volt (keV) which is 1000 electron volts. Energy is also measured in joules, and $1\,eV \equiv 1.60219 \times 10^{-19}$ J (coulomb-volts). Energy and wavelength are related in the fundamental equation

$$E = h\nu = \frac{hc}{\lambda}$$

where E is the energy, λ the wavelength, ν the frequency, c the velocity of light and h is Planck's constant. Thus the energy equivalent of a 1 Å X-ray photon would be

$$E = \frac{6.626 \times 10^{-34} \times 3 \times 10^8}{10^{-10}}\ J$$

$$= 2 \times 10^{-15}\ J$$

or

$$\frac{2 \times 10^{-15}}{1.6 \times 10^{-9}}\ eV$$

$$= 12.4\ keV$$

This gives a very useful relationship for the conversion of wavelength to energy, i.e.

$$\lambda(\text{Å}) = \frac{12.4}{E(\text{keV})} \qquad (2\text{-}1)$$

TABLE 2-1.
Wavelength standards

Primary		Secondary	
W Kα$_1$	λ = 0.2090100 Å	Ag Kα$_1$	λ = 0.5594075 Å $\pm$ 0.7 × 10^{-6}
		Mo Kα$_1$	λ = 0.709300 Å $\pm$ 0.9 × 10^{-6}
		Cu Kα$_1$	λ = 1.540562 Å $\pm$ 1.8 × 10^{-6}
		Cr Kα$_2$	λ = 2.293606 Å $\pm$ 3.0 × 10^{-6}

Due to uncertainties in the values of h and c,[4] there is some inconsistency between the wavelength and energy scales. The following relationship holds

$$E(\text{keV}) = \frac{12.3964}{\lambda(\text{Å})} = \frac{12.3964}{1.00202 \times \lambda(\text{kXU})}$$

However, for identification purposes in routine analytical X-ray spectrometry accuracies of the order of 1 in 1000 are generally quite sufficient. Detailed tables of more accurate data are available[5] where these are required.

2.2 ELECTRON CONFIGURATIONS OF THE ELEMENTS

Atoms are made up of nuclei surrounded by electrons, the number of electrons being equal to the atomic number (i.e. the number of nuclear protons) of the atom. Thus magnesium, atomic number 12, has 12 electrons; barium, atomic number 56, has 56 electrons and so on. The configuration of the electrons within an atom follows a definite pattern and simple rules can be applied to predict their states. Each electron represents a certain amount of energy and this energy can be described by four parameters, the so-called "quantum numbers" of the electron. These quantum numbers are n, the principal quantum number, which can take positive integral values 1, 2, 3, 4, etc.; l, the angular quantum number, which can have values of 0 through $(n - 1)$; m, the magnetic quantum number, which can have values of $-l$ through 0 through $+l$; and s, the spin quantum number, which has values of $\pm 1/2$.

The Pauli exclusion principle states that no two electrons within an atom can have the same set of quantum numbers, thus there can only be two electrons in the first principal sub-shell where $n = 1$. Similarly, there are eight combinations for $n = 2$, 18 for $n = 3$, etc. In general, there will be $2n^2$ possible combinations. Table 2-2 shows the orbitals of which the principal sub-shells are made up, and it will be seen that where $l = 0$, the orbital is called an s (sharp) orbital, where $l = 1$ a p (principal) orbital; where $l = 2$ a d (diffuse) orbital; and where $l = 3$ an f (fundamental) orbital. The names in parentheses are the spectroscopic terms for the orbitals.

TABLE 2-2.
Electron configurations in the first four principal sub-shells

Shell	n	l	m	n	Orbital	Maximum number of electrons
K	1	0	0	$\pm 1/2$	1s	2
L	2	0	0	$\pm 1/2$	2s	
		1	+1	$\pm 1/2$	2p	8
		1	0	$\pm 1/2$		
		1	−1	$\pm 1/2$		
M	3	0	0	$\pm 1/2$	3s	
		1	+1	$\pm 1/2$	3p	
		1	0	$\pm 1/2$		
		1	−1	$\pm 1/2$		
		2	+2	$\pm 1/2$		18
		2	+1	$\pm 1/2$		
		2	0	$\pm 1/2$	3d	
		2	−1	$\pm 1/2$		
		2	−2	$\pm 1/2$		
N	4	0	0	$\pm 1/2$	4s	
		1	+1	$\pm 1/2$	4p	
		1	0	$\pm 1/2$		
		1	−1	$\pm 1/2$		
		2	+2	$\pm 1/2$		
		2	+1	$\pm 1/2$		
		2	0	$\pm 1/2$	4d	
		2	−1	$\pm 1/2$		
		2	−2	$\pm 1/2$		32
		3	+3	$\pm 1/2$		
		3	+2	$\pm 1/2$		
		3	+1	$\pm 1/2$		
		3	0	$\pm 1/2$	4f	
		3	−1	$\pm 1/2$		
		3	−2	$\pm 1/2$		
		3	−3	$\pm 1/2$		

Table 2-3 shows the actual electron configurations for all of the elements in the periodic classification and it will be seen that, with certain exceptions, elements of increasing atomic number are built up by a regular sequence of electron additions in the order 1s 2s 2p 3s 3p 4s 3d 4p, etc.

A good deal of interaction is possible between the electrons, particularly where one is considering the energy required to displace an electron from its normal or "ground state" configuration. One of the most important of these synergistic effects is described by the coupling of the l and s quantum numbers, the result being a vector sum of the **l** and **s** moments. Thus,

$$j = \mathbf{l} + \mathbf{s} \tag{2-2}$$

where j is the total moment.

TABLE 2-3†

Period	Element	Electronic configuration								K Binding energies (keV)
1	1 H	1s								0.014
	2 He	1s²								0.025
2	3 Li	K	2s							0.055
	4 Be	K	2s²							0.011
	5 B	K	2s²	2p						0.181
	6 C	K	2s²	2p²						0.284
	7 N	K	2s²	2p³						0.399
	8 O	K	2s²	2p⁴						0.532
	9 F	K	2s²	2p⁵						0.686
	10 Ne	K	2s²	2p⁶						0.867
3	11 Na	K	L	3s						1.07
	12 Mg	K	L	3s²						1.31
	13 Al	K	L	3s²	3p					1.56
	14 Si	K	L	3s²	3p²					1.84
	15 P	K	L	3s²	3p³					2.15
	16 S	K	L	3s²	3p⁴					2.47
	17 Cl	K	L	3s²	3p⁵					2.82
	18 Ar	K	L	3s²	3p⁶					3.20
4	19 K	K	L	3s²	3p⁶		4s			3.61
	20 Ca	K	L	3s²	3p⁶		4s²			4.04
	21 Sc	K	L	3s²	3p⁶	3d¹	4s²			4.49
	22 Ti	K	L	3s²	3p⁶	3d²	4s²			4.97
	23 V	K	L	3s²	3p⁶	3d³	4s²			5.47
	24 Cr	K	L	3s²	3p⁶	3d⁴	4s²			5.99
	25 Mn	K	L	3s²	3p⁶	3d⁵	4s²			6.54
	26 Fe	K	L	3s²	3p⁶	3d⁶	4s²			7.11
	27 Co	K	L	3s²	3p⁶	3d⁷	4s²			7.71
	28 Ni	K	L	3s²	3p⁶	3d⁸	4s²			8.33
	29 Cu	K	L	3s²	3p⁶	3d⁹	4s²			8.98
	30 Zn	K	L	3s²	3p⁶	3d¹⁰	4s²			9.66
	31 Ga	K	L	M	4s²	4p				10.37
	32 Ge	K	L	M	4s²	4p²				11.10
	33 As	K	L	M	4s²	4p³				11.87
	34 Se	K	L	M	4s²	4p⁴				12.66
	35 Br	K	L	M	4s²	4p⁵				13.47
	36 Kr	K	L	M	4s²	4p⁶				14.33
5	37 Rb	K	L	M	4s²	4p⁶		5s		15.20
	38 Sr	K	L	M	4s²	4p⁶		5s²		16.11
	39 Y	K	L	M	4s²	4p⁶	4d	5s²		17.04
	40 Zr	K	L	M	4s²	4p⁶	4d²	5s²		18.00
	41 Nb	K	L	M	4s²	4p⁶	4d⁴	5s		18.99
	42 Mo	K	L	M	4s²	4p⁶	4d⁵	5s		20.00
	43 Tc	K	L	M	4s²	4p⁶	4d⁵	5s²		21.04
	44 Ru	K	L	M	4s²	4p⁶	4d⁷	5s		22.12
	45 Rh	K	L	M	4s²	4p⁶	4d⁸	5s		23.22
	46 Pd	K	L	M	4s²	4p⁶	4d¹⁰			24.35
	47 Ag	K	L	M	4s²	4p⁶	4d¹⁰	5s		25.51
	48 Cd	K	L	M	4s²	4p⁶	4d¹⁰	5s²		26.71
	49 In	K	L	M	4s²	4p⁶	4d¹⁰	5s²	5p	27.94
	50 Sn	K	L	M	4s²	4p⁶	4d¹⁰	5s²	5p²	29.20

† K, L, M, and N represent completely full levels.

Period	Element					Electronic configuration								K Binding energies (keV)
	51 Sb	K	L	M	$4s^2$	$4p^6$	$4d^{10}$		$5s^2$	$5p^3$				30.49
	52 Te	K	L	M	$4s^2$	$4p^6$	$4d^{10}$		$5s^2$	$5p^4$				31.81
	53 I	K	L	M	$4s^2$	$4p^6$	$4d^{10}$		$5s^2$	$5p^5$				33.17
	54 Xe	K	L	M	$4s^2$	$4p^6$	$4d^{10}$		$5s^2$	$5p^6$				34.56
6	55 Cs	K	L	M	$4s^2$	$4p^6$	$4d^{10}$		$5s^2$	$5p^6$		$6s$		35.99
	56 Ba	K	L	M	$4s^2$	$4p^6$	$4d^{10}$		$5s^2$	$5p^6$		$6s^2$		37.44
	57 La	K	L	M	$4s^2$	$4p^6$	$4d^{10}$		$5s^2$	$5p^6$	$5d$	$6s^2$		38.93
	58 Ce	K	L	M	$4s^2$	$4p^6$	$4d^{10}$	$4f^2$	$5s^2$	$5p^6$		$6s^2$		40.44
	59 Pr	K	L	M	$4s^2$	$4p^6$	$4d^{10}$	$4f^3$	$5s^2$	$5p^6$		$6s^2$		41.99
	60 Nd	K	L	M	$4s^2$	$4p^6$	$4d^{10}$	$4f^4$	$5s^2$	$5p^6$		$6s^2$		43.57
	61 Pm	K	L	M	$4s^2$	$4p^6$	$4d^{10}$	$4f^5$	$5s^2$	$5p^6$		$6s^2$		45.19
	62 Sm	K	L	M	$4s^2$	$4p^6$	$4d^{10}$	$4f^6$	$5s^2$	$5p^6$		$6s^2$		46.84
	63 Eu	K	L	M	$4s^2$	$4p^6$	$4d^{10}$	$4f^7$	$5s^2$	$5p^6$		$6s^2$		48.52
	64 Gd	K	L	M	$4s^2$	$4p^6$	$4d^{10}$	$4f^7$	$5s^2$	$5p^6$	$5d$	$6s^2$		50.24
	65 Tb	K	L	M	$4s^2$	$4p^6$	$4d^{10}$	$4f^9$	$5s^2$	$5p^6$		$6s^2$		52.00
	66 Dy	K	L	M	$4s^2$	$4p^6$	$4d^{10}$	$4f^{10}$	$5s^2$	$5p^6$		$6s^2$		53.79
	67 Ho	K	L	M	$4s^2$	$4p^6$	$4d^{10}$	$4f^{11}$	$5s^2$	$5p^6$		$6s^2$		55.62
	68 Er	K	L	M	$4s^2$	$4p^6$	$4d^{10}$	$4f^{12}$	$5s^2$	$5p^6$		$6s^2$		57.49
	69 Tm	K	L	M	$4s^2$	$4p^6$	$4d^{10}$	$4f^{13}$	$5s^2$	$5p^6$		$6s^2$		59.39
	70 Yb	K	L	M	$4s^2$	$4p^6$	$4d^{10}$	$4f^{14}$	$5s^6$	$5p^6$		$6s^2$		61.33
	71 Lu	K	L	M	N	$5s^2$	$5p^6$	$5d$	$6s^2$					63.31
	72 Hf	K	L	M	N	$5s^2$	$5p^6$	$5d^2$	$6s^2$					65.35
	73 Ta	K	L	M	N	$5s^2$	$5p^6$	$5d^3$	$6s^2$					67.42
	74 W	K	L	M	N	$5s^2$	$5p^6$	$5d^4$	$6s^2$					69.53
	75 Re	K	L	M	N	$5s^2$	$5p^6$	$5d^5$	$6s^2$					71.68
	76 Os	K	L	M	N	$5s^2$	$5p^6$	$5d^6$	$6s^2$					73.87
	77 Ir	K	L	M	N	$5s^2$	$5p^6$	$5d^7$	$6s^2$					76.11
	78 Pt	K	L	M	N	$5s^2$	$5p^6$	$5d^9$	$6s$					78.40
	79 Au	K	L	M	N	$5s^2$	$5p^6$	$5d^{10}$	$6s$					80.73
	80 Hg	K	L	M	N	$5s^1$	$5p^6$	$5d^{10}$	$6s^2$					83.10
	81 Tl	K	L	M	N	$5s^2$	$5p^6$	$5d^{10}$	$6s^2$	$6p$				85.53
	82 Pb	K	L	M	N	$5s^2$	$5p^6$	$5d^{10}$	$6s^2$	$6p^2$				88.01
	83 Bi	K	L	M	N	$5s^2$	$5p^6$	$5d^{10}$	$6s^2$	$6p^3$				90.53
	84 Po	K	L	M	N	$5s^2$	$5p^6$	$5d^{10}$	$6s^2$	$6p^4$				93.11
	85 At	K	L	M	N	$5s^2$	$5p^6$	$5d^{10}$	$6s^2$	$6p^5$				95.73
	86 Rn	K	L	M	N	$5s^2$	$5p^6$	$5d^{10}$	$6s^2$	$6p^6$				98.40
7	87 Fr	K	L	M	N	$5s^2$	$5p^6$	$5d^{10}$		$6s^2$	$6p^6$		$7s$	101.14
	88 Ra	K	L	M	N	$5s^2$	$5p^6$	$5d^{10}$		$6s^2$	$6p^6$		$7s^2$	103.92
	89 Ac	K	L	M	N	$5s^2$	$5p^6$	$5d^{10}$		$6s^2$	$6p^6$	$6d$	$7s^2$	106.76
	90 Th	K	L	M	N	$5s^2$	$5p^6$	$5d^{10}$		$6s^2$	$6p^6$	$6d^2$	$7s^2$	109.65
	91 Pa	K	L	M	N	$5s^2$	$5p^6$	$5d^{10}$	$5f^2$	$6s^2$	$6p^6$	$6d$	$7s^2$	112.60
	92 U	K	L	M	N	$5s^2$	$5p^6$	$5d^{10}$	$5f^3$	$6s^2$	$6p^6$	$6d$	$7s^2$	115.61
	93 Np	K	L	M	N	$5s^2$	$5p^6$	$5d^{10}$	$5f^4$	$6s^2$	$6p^6$	$6d$	$7s^2$	118.68
	94 Pu	K	L	M	N	$5s^2$	$5p^6$	$5d^{10}$	$5f^6$	$6s^2$	$6p^6$		$7s^2$	121.82
	95 Am	K	L	M	N	$5s^2$	$5p^6$	$5d^{10}$	$5f^7$	$6s^2$	$6p^6$		$7s^2$	125.03
	96 Cm	K	L	M	N	$5s^2$	$5p^6$	$5d^{10}$	$5f^7$	$6s^2$	$6p^6$	$6d$	$7s^2$	128.22
	97 Bk	K	L	M	N	$5s^2$	$5p^6$	$5d^{10}$	$5f^8$	$6s^2$	$6p^6$	$6d$	$7s^2$	131.59
	98 Cf	K	L	M	N	$5s^2$	$5p^6$	$5d^{10}$	$5f^{10}$	$6s^2$	$6p^6$		$7s^2$	135.96
	99 Es	K	L	M	N	$5s^2$	$5p^6$	$5d^{10}$	$5f^{11}$	$6s^2$	$6p^6$		$7s^2$	139.49
	100 Fm	K	L	M	N	$5s^2$	$5p^6$	$5d^{10}$	$5f^{12}$	$6s^2$	$6p^6$		$7s^2$	143.09
	101 Md	K	L	M	N	$5s^2$	$5p^6$	$5d^{10}$	$5f^{13}$	$6s^2$	$6p^6$		$7s^2$	146.78
	102 No	K	L	M	N	$5s^2$	$5p^6$	$5d^{10}$	$5f^{14}$	$6s^2$	$6p^6$		$7s^2$	150.54
	103 Lw	K	L	M	N	$5s^2$	$5p^6$	$5d^{10}$	$5f^{14}$	$6s^2$	$6p^6$	$6d$	$7s^2$	154.38

TABLE 2-4.

Transition level	l	j	Transition level	l	j
K	0	1/2	N_I	0	1/2
			N_{II}	1	1/2
L_I	0	1/2	N_{III}	1	3/2
L_{II}	1	1/2	N_{IV}	2	3/2
L_{III}	1	3/2	N_V	2	5/2
			N_{VI}	3	5/2
M_I	0	1/2	N_{VII}	3	7/2
M_{II}	1	1/2			
M_{III}	1	3/2			
M_{IV}	2	3/2			
M_V	2	5/2			

The combination of these quantum numbers gives rise to the so-called transition levels. Table 2-4 lists these levels and the possible values of j. It will be seen that there are a single K level, 3 L levels, 5 M levels and 7 N levels.

2.3 PRODUCTION OF CONTINUOUS RADIATION

If an element is bombarded with high energy electrons, there will be interaction between the electrons making up the atoms of the element and the bombarding electrons. Any interaction will involve energy changes within the atom and the atom will later return to its original (lowest) energy state by giving up any extra energy by the emission of radiation. Among the types of interaction which may occur are scattering of the bombarding electrons, increased perturbation of the electrons of the atom and, in the extreme case, removal of these electrons from their original atomic sites. This last effect may give rise to the production of characteristic radiation from the atom and this effect will be discussed in the next sections. Increasing the perturbation of the atomic electrons without removing them completely from the atom, a process called excitation (as opposed to ionization which involves *complete* removal of the electron), may itself be of two types. The first of these involves the promotion of an electron to an unfilled orbital from whence it returns to its original position with the emission of radiation—in this case, radiation in the ultraviolet region. This first effect is generally restricted to the loosely bound outer electrons. The second effect involves the more tightly bound (i.e., less easily removed) inner electrons which can be made to oscillate, i.e., to accept energy without actually leaving their site positions. Both of these effects result in loss of energy of the primary incident electron.

The atom itself will emit many forms of radiation which may include infrared, ultraviolet and X-radiation of which the last is of immediate concern. Figure 2-1

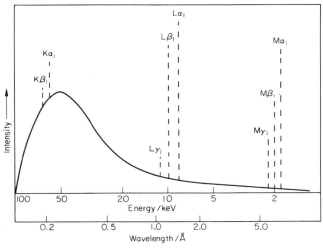

Fig. 2-1   Intensity distribution from a tungsten anode X-ray tube at 100 kV

shows the intensity distribution obtained from a tungsten target X-ray tube and it consists of a broad band of continuous or white radiation, superimposed on top of which are characteristic wavelengths. The distribution of radiation of the continuum can be expressed in terms of Kramer's formula.

$$I(\lambda)\,d\lambda = KiZ\left[\frac{\lambda}{\lambda_{min}} - 1\right]\frac{1}{\lambda^2}\,d\lambda \qquad (2\text{-}3)$$

It will be seen from this expression that the intensity distribution is a linear function of the X-ray tube current i and the atomic number Z of the X-ray tube anode material. λ_{min} represents the minimum wavelength of the continuum and this is found to correspond to the maximum energy of the electrons, i.e., the operating voltage V_0 of the X-ray tube. In fact

$$\lambda_{min} = \frac{hc}{V_0} = \frac{12.4}{V_0} \qquad (2\text{-}4)$$

The spectrum shown in Fig. 2-1 was obtained at 100 kV, hence in this instance, $\lambda_{min} = 0.124$ Å.

The intensity maximum of the continuum can be obtained by differentiating the Kramer formula, viz.

$$I = K'\left[\frac{1}{\lambda \cdot \lambda_{min}} - \frac{1}{\lambda^2}\right]$$

and again

$$\frac{dI}{d\lambda} = -\left[\frac{K'}{\lambda^2 \cdot \lambda_{min}} + \frac{2K'}{\lambda^3}\right] = 0$$

or

$$\frac{1}{\lambda_{min}} = \frac{2}{\lambda}$$

i.e.

$$\lambda_{max} = 2\lambda_{min}$$

Hence the intensity maximum of the continuum always occurs at twice the value of the minimum wavelength.

In practice, the continuum from an X-ray tube is considerably modified by self absorption of the anode, Compton scatter of the continuum and absorption by the X-ray tube window. However, the general shape shown in Fig. 2-1 is usually maintained.

2.4 EXCITATION AND DE-EXCITATION

The process of the displacement of an electron from its normal or "ground" state is called excitation. Figure 2-2(a) illustrates the removal of a K level electron by a primary photon having an energy equal to $h\nu - \phi_K$ where ϕ_K is the binding energy of the K electron, leaving the atom in the K^+ state.

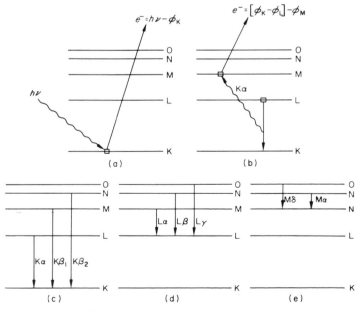

Fig. 2-2   Excitation and de-excitation

The atom can return to the non-excited or "ground state" by various processes of which two are dominant. The first of these is where an electron from one of the upper levels (i.e., a level with a value of n larger than that of the ionized level) falls to the excited level. As will be seen shortly, such an electron transference will always be accompanied by the emission of radiation. The second process is similar in its initial stage to the first, but in this case, the radiation produced following the transference of an electron from an upper level is not emitted, but gives up its energy by further ionization within the atom. This process is known as the Auger process[6] and is illustrated in Fig. 2-2(b). Here the initial excitation stage produced a K vacancy and this was followed by a de-excitation involving transference of an L electron. Such a transfer results in the production of $K\alpha$ radiation, but in this instance the photon of radiation does not leave the atom but itself ejects an M electron. This effect is called "autoionization" and is very akin to the phenomena found in the excitation observed in optical spectra. As in the initial excitation, this photoelectron will have an energy equal to that of the incident radiation (in this case $\phi_K - \phi_L$) less that required for the ejection of the electron, i.e., the binding energy of the ejected electron.

Note that the Auger process can lead to the production of two (or more) vacancies in the upper levels—in this instance L^+ and M^+. As will be seen later, this double vacancy production is an important factor in the formation of so-called satellite lines.

The quantity of radiation from a certain level will be dependent upon the relative efficiency of the two opposing de-excitation processes involved. This relative efficiency is usually expressed in terms of the fluorescent yield. The fluorescent yield ω is defined as the number n of X-ray photons emitted within a given series, divided by the total number N of vacancies formed in the associated level, each with the same time increment. A series represents a set of wavelengths all of which arise from transitions to the same atomic sub-shell. In general terms for the K series

$$\omega_K = \frac{\sum (n)_K}{N_K} = \frac{n(K\alpha_1) + n(K\alpha_2) + n(K\beta_1) + n(K\beta_2)\dots \text{etc.}}{N_K} \qquad (2\text{-}5)$$

For example, if within a given period of time, 80 vacancies were formed, and 30, 15, 5 and 1 X-ray photons were emitted corresponding to $K\alpha_1 K\alpha_2 K\beta_1 K\beta_2$ respectively, the K fluorescent yield would be

$$\omega_K = \frac{30 + 15 + 5 + 1}{80} = \frac{51}{80} = 0.638.$$

Note that the fraction of vacancies filled via the Auger process is always one minus the fluorescent yield or, in the given example, 0.362, i.e., 29 vacancies.

Figure 2-3 shows the correlation between K and L fluorescence yield as a function of atomic number. It will be seen that the fluorescent yield decreases markedly with decrease of atomic number, both for the K and L series. One important experimental point to note is that whereas the difference in K and L fluorescent yield for a given element may be as much as 5 to 10, the difference in K and L fluorescent yield for a given wavelength is generally only around 1.5 to 2. As an example, Zn Kα and W Lα both have wavelengths of around 1.47 Å and the respective equivalent fluorescent yields are 0.5 and 0.3.

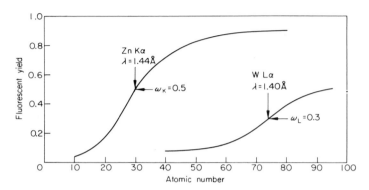

Fig. 2-3   Fluorescent yield as a function of atomic number

There is an increasing demand for accurate fluorescent yield data,[7] particularly with the growing interest in the so-called "fundamental methods" of quantitative X-ray spectrometry. These fundamental methods equate elemental concentration and the intensities of selected characteristic X-ray emission wavelengths of a specimen, in terms of certain fundamental constants such as mass absorption coefficients, excitation probabilities, fluorescent yields, and so on. The great benefit to be gained by use of such algorithms is the ability to work without large numbers of calibration standards, but in turn, sufficiently accurate fundamental constants are required in order to give acceptable analytical accuracy (see p. 123). K fluorescent yield values are fairly well established and are generally measurable to accuracies of the order of 2–4%.[8,9] Unfortunately, it is far more difficult to obtain accurate data for the L fluorescent yields owing to the complications of vacancy distributions between the various sub-shells.

2.5 IONIZATION AND EXCITATION POTENTIALS

When an inner shell electron is removed from an atom, the atom is said to be ionized in the shell from which the electron was removed. For example, if one of the K electrons were ejected, the atom would be in the K^+ state, if an L electron were removed, the atom would be in the L^+ state, and so on. Where two electrons are removed simultaneously, the atom is said to be doubly ionized, normally indicated by KL for ionization in both K and L shells, LL for double ionization in the L shell, etc. The energy required to remove an electron is called the binding energy, which is in effect the energy required to remove the electron beyond the Fermi level of the atom. The value of the binding energy varies from atom to atom and from shell to shell within an atom, since it is dependent upon the angular and spin quantum numbers of an electron within a given shell. Table 2-3 shows the binding energies for the K level of each element.

Removal of the first (1s) electron would lead to the K^+ state, and the atom can return to the unexcited state, the so-called ground state, by de-excitation via a series of electron transitions (see Fig. 2-2). For example, transference of an electron from the L level to the K level involves a change in the state of the atom from K^+ to L^+. A similar transference of an M electron to the L level would further change the state from L^+ to M^+. This procedure may be continued until the atom regains the ground state—the last electron transferred coming from the low energy extremities of the atom where electron changes are almost continuous (e.g., the conduction or valence bands). This is obviously only one of many possible ways of de-excitation, for instance, in the given example, it would also be possible to de-excite by a single transference of an outer electron to the K level.

It will be seen by examination of one of the de-excitation steps that a significant energy change may be involved. For instance, the transference of an L electron to the K level in a copper atom in the K^+ initial state would involve an energy change of about 8040 eV. The binding energy in the copper K level is 8973 eV and that for the L level 933 eV, i.e., 8973 eV are required to eject an electron from the K level to give the K^+ state, and 933 eV are required to eject an electron from the L level to give the L^+ state. In moving from K^+ to L^+, the total energy change will be $8973 - 933 = 8040$ eV. This extra energy is emitted during electron transference, i.e., during de-excitation, and is emitted as an X-ray photon with an energy of 8040 eV.

In practice, millions of atoms are involved in the excitation of a given specimen and all possible de-excitation routes will be taken. What will differ from element to element, however, is the distribution of the total de-excitation over the various routes.

The various de-excitation routes can be defined by fairly simple selection rules which account for the majority of observed wavelengths. These are generally classified as one of three types.

(a) Normal transitions—which can be defined by a simple set of selection rules.

(b) Forbidden transitions—observed wavelengths which do not obey the same set of selection rules.

(c) Satellite lines—which are mainly wavelengths occurring from dual ionizations.

2.6   TRANSITIONS AND DIAGRAM LINES

The normal transitions (diagram lines) are defined by three simple atomic selection rules, viz.

$$\Delta n \geqslant 1$$
$$\Delta l = \pm 1$$
$$\Delta j = \pm 1 \text{ or } 0$$

Unfortunately, the nomenclature of the associated X-ray wavelengths is archiac and unsystematic, but several broad generalizations can be made. For instance, the final resting place of the transferred electron *always* determines the series of the associated radiation. Thus, ionization in the K shell followed by the filling of the K vacancy leads to the production of K series radiation. The filling of an L vacancy gives rise to L series radiation, and so on. Further, the strongest line in a given series is called the α line and weaker lines are called β, γ, δ, and so on, although the relative intensities of these lines bear little resemblance to the sequence of labelling.

Study of these selection rules will indicate that a given transition, e.g., $2p \rightarrow 1s$, can give rise to more than one line since the 2p electron being transferred has two states depending upon the sign of the spin quantum number. Where the spin is $-\frac{1}{2}$, the j quantum number will be $+\frac{1}{2}$ (equation (2-2)) since for a p electron $l = 1$. The spectroscopic nomenclature in this case is $^2P_{1/2}$. Similarly, if the spin is $+\frac{1}{2}$, j will equal 3/2 and the spectroscopic nomenclature is $^2P_{3/2}$. Thus, two Kα lines occur, the Kα_1 and the Kα_2.

Since both l and j are used to define the state of an electron within a given shell, it is possible to construct so-called "levels" between which electrons are transferred. Table 2-4 shows the construction of these levels and it will be seen that the number of levels is always $2n - 1$, thus 1 K level, 3 L levels, 5 M levels, etc. Note that only positive values of j are possible, hence only a single j value arises from the state $l = 0$. This is because the s orbital (when $l = 0$, the orbital is s) is spherically symmetrical and therefore unpolarizable.

It is often confusing for the newcomer to the field of X-ray spectroscopy to find that there are no less than three different ways of expressing a transition. For example, the Kα_1 and Kα_2 lines already mentioned could be explained thus:

$$K\alpha_1 : L_{III} \rightarrow K \quad \text{or} \quad 2p^{3/2} \rightarrow 1s \quad \text{or} \quad {}^2P_{3/2} \rightarrow {}^1S_0$$
$$K\alpha_2 : L_{II} \rightarrow K \quad \text{or} \quad 2p^{1/2} \rightarrow 1s \quad \text{or} \quad {}^2P_{1/2} \rightarrow {}^1S_0$$

The first mentioned will be found in most textbooks and tabular values of wavelengths. Although it is at first sight the most convenient method to use, it is of little use in the discussion of satellite lines and forbidden transitions. The second method will be generally familiar to the chemist and has the great merit of indicating exactly which electron states are involved in a given transition. The third scheme is simply the correct spectroscopic nomenclature for the previous method and indicates the electron states. The last method is generally not favoured by the analytical spectroscopist and further description in this text will utilize mainly the second scheme.

2.7 SATELLITE LINES

Satellite lines were observed even by the very early workers in the field of X-ray spectroscopy and much of the theory for their production is well established.[10–12]

Satellite lines are lines which occur following dual ionization of the atom, that is to say, a second ionization occurs within the lifetime of the first excited state. Any wavelength emission following transfer of an electron to a vacancy within an atom where a second vacancy exists will differ from that produced where the second vacancy does not exist. This will be apparent, since the second vacancy must cause some perturbation and a general increase in the energy levels of the atom. Thus, the wavelengths associated with dual vacancies will be shorter than their single vacancy counterparts. Dual ionization can take place by various means and one of these, the Auger process of auto-ionization, has already been discussed. The Auger process is indeed the dominant process for the production of double vacancies in all levels above the L levels and it is due to this process that the majority of satellites in the M, L, N, etc., series lines occur. However, this would not be the reason for dual ionization involving K and L levels, since the Auger process begins following the internal conversion of an X-ray photon which will in this instance have itself come from transference between the two levels in question. It seems certain, therefore, that K series satellites exist owing to double ionization by single electron impacts.

2.8 CHARACTERISTIC LINE SPECTRA

The types of transitions allowable are best illustrated by means of a series of transition diagrams and their appropriate emission spectra. The following can be taken as typical examples for most of the analytical wavelength region.

2.9 K SPECTRA

K spectra arise following the transference of electrons to K shell vacancies. K Spectra are relatively simple and generally consist of two doublets with an extra line occurring for the higher atomic number elements.

2.9.1 The tin K spectrum

Tin is atomic number 50 and has, in the ground state, filled K, L and M levels, filled 4s, 4p, 4d and 5s levels, plus two electrons in the 5p level. Figure 2-4 shows

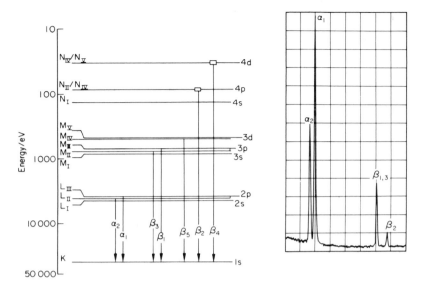

Fig. 2-4   The K emission spectrum of tin

the energy level diagram and indicates the approximate binding energies of the various sub-shells.[13] Following ejection of a K (1s) electron, seven lines may be observed. According to the selection rules, only six lines should be allowable, in fact 3 doublets corresponding to transitions from the 4p, 3p and 2p levels. These lines do indeed occur; however, only the α_1 and α_2 appear on the emission diagram as a recognizable doublet. This is because the energy required to polarize the p orbitals becomes smaller with decrease of electron density. For example, in the case of tin, the difference between the states $2p^{3/2}$ and $2p^{1/2}$ is 227 eV and between the states $3p^{3/2}$ and $3p^{1/2}$ only 42 eV. The visual effect on the emission spectrum is readily seen since the angular dispersion of a spectrometer is directly proportional to the absolute value of the energy difference between two lines (p. 76). Thus, whereas the α_1 and α_2 are clearly seen as two sharp and well separated lines, the β_1/β_3 doublet is completely unresolved. This is even more so in the case of the 4p → 1s transition which is generally referred to not as a doublet at all but simply the β_2. Two lines do in fact exist, however, and these are identified by β_2' and β_2''.

Two additional transitions occur in the tin K spectrum and these are the forbidden transitions 3d → 1s and 4d → 1s. These transitions are classified as forbidden since they both represent a Δl of 2 and hence disobey the normal selection rules. Like all forbidden lines, they are very weak and are not generally visible in the measured spectrum.

A further point of interest is that the transition (doublet) 5p → 1s would be allowed, but in the ground state, tin has just two electrons in the 5p level.

Chemical bonding may, of course, increase this number to the full compliment of six. A very weak line corresponding to this transition (29.195 keV) is indeed observed in certain tin compounds.

2.9.2 The copper K spectrum

Copper is atomic number 29 and has filled K and L levels, filled 3s, 3p and 4s levels and an almost full 3d level. Thus, unlike tin, it has no electrons in the 4p level while in the ground state, although there might be some partial filling of this level during chemical bonding. Figure 2-5 shows the effect of the unfilled 4p level since the β_2 line is absent. The absence of the β_2 line does, however, allow one to see the very weak β_5 line which comes from the forbidden 3d → 1s transition.

A further difference to be noted between the copper and tin spectra is the lower degree of splitting of the α_1 and α_2 doublet in the case of copper. This is because the absolute energy difference between α_1 and α_2 in copper is almost an order of magnitude less than in tin.

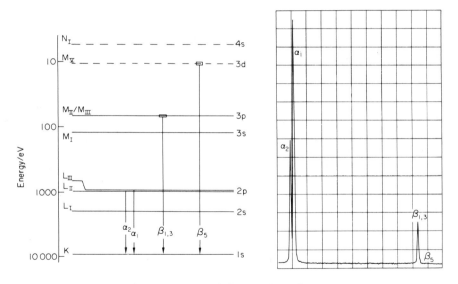

Fig. 2-5   The K emission spectrum of copper

2.9.3 The calcium K spectrum

Calcium is atomic number 20 and in the ground state has filled K, L, 3s, 3p, and 4s levels. The 3d level is unoccupied, hence the forbidden 3d → 1s transition is not observed (see Fig. 2-6). Also in this instance, the energy gap between $2p^{3/2}$ and $2p^{1/2}$ is so small that there is no observable splitting of α_1, α_2.

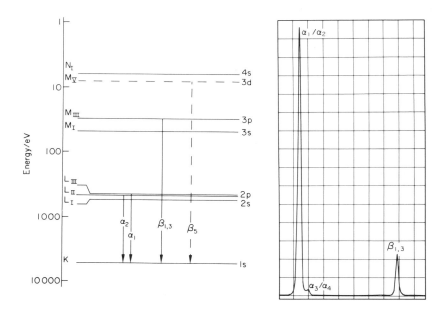

Fig. 2-6   The K emission spectrum of calcium

Several other interesting features also begin to appear at this level of orbital occupancy and these involve the satellite lines to be seen to the high energy (right hand side of the figure) side of the $\alpha_1\alpha_2$, $\beta_1\beta_3$ doublets. These satellite lines become more and more pronounced as the atomic number (i.e., the number of electrons) decreases.

2.9.4 The aluminium K spectrum

Aluminium is atomic number 13 and in the ground state has filled K, L and 3s levels and a single 3p electron. Since the $\beta_{1,3}$ line comes from a 3p transition, only a very weak broad line is observed (Fig. 2-7) and the shape of the line is very dependent upon the chemical bonding of the aluminium atoms. Figure 2-8, taken from work by Fischer,[14] clearly indicates the effect of bonding and illustrates the reason for the band-like nature of the line which now arises not from the discrete energy level characteristic of an atomic orbital, but rather from the energy band of the molecular orbital of aluminium and oxygen. This change from line to band spectra is very typical of this part of the atomic number region. Also shown on the diagram is the satellite K'_β line.

Returning for the moment to Fig. 2-7, two satellite doublets α_3, α_4 and α_5, α_6 are to be seen to the high energy side of the α_1, α_2 doublet. Figure 2-9 shows an enlarged view of this portion of the spectrum and the associated energy level

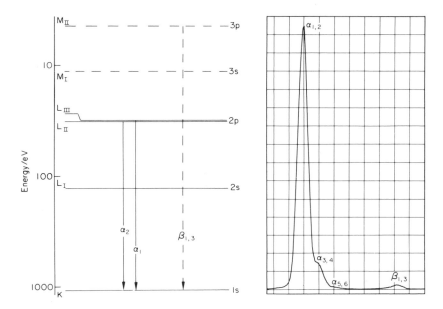

Fig. 2-7 The K emission spectrum of aluminium

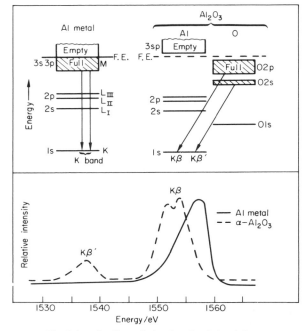

Fig. 2-8 The K_β emission bands of aluminium

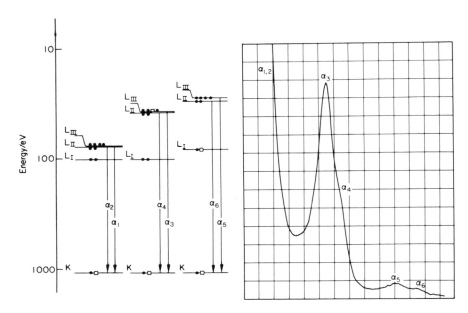

Fig. 2-9   The K satellite lines of aluminium

diagram indicates that both satellite doublets occur from double ionizations. It has already been mentioned that removal of two electrons from the same atom is possible and this is most likely to occur where the electron density is very low, i.e., in the case of the lower atomic number elements. Two possible double ionizations are possible, namely KL_{III} and KL_I (KL_{II} would of course be the same as KL_{III}). In either case, removal of the 2p (or 2s) electron increases the distance and hence the energy gap between 2p and 1s, with the resulting production of two extra satellite pairs. (In fact, a fifth $K\alpha$ satellite is observed between the $K\alpha_2$ and the $K\alpha_3$. This line is the $K\alpha'$ which arises from a more complex dual ionization.)

2.9.5   The oxygen K spectrum

Oxygen is atomic number eight and only the 1s and 2s levels are filled. In the ground state, four 2p electrons are present but these will obviously be directly involved in bonding. Figure 2-10 illustrates the oxygen K emission spectrum obtained from $LiOH.H_2O$ and clearly shows the band-like nature of the line. Two maxima are seen in the spectrum and these correspond roughly to the upper and lower levels of the 2p orbital band.

Oxygen has no 3p electrons and therefore gives no β lines.

The oxygen K spectrum is typical of the K spectra of the very low atomic number elements. Even lithium, which is atomic number three, and has the

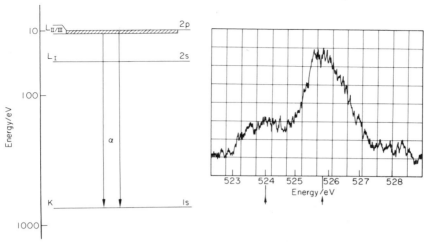

Fig. 2-10   The K emission spectrum of oxygen

structure $1s^2 2s^1$ gives a K spectrum. At first sight, this would seem impossible since there is no 2p character in the lithium atom; however, it is possible for electrons to be transferred to the 1s vacancy from a molecular orbital arising from a ligand between two dissimilar atoms (e.g., in lithium compounds) or even between two similar atoms.

The possible use of the shape and distribution of X-ray emission bands is presently providing considerable insight into the mechanism of chemical bonding.[15–17] This subject will be discussed further in Chapter 8, p. 141.

2.10   L SPECTRA

L Spectra arise following the transference of electrons to fill vacancies in the L levels. Since there are three L levels, compared with only a single K level, there will be a far greater number of possible L transitions allowable within the selection rules. L Spectra are thus more complex than K spectra and between 20 and 30 normal (diagram) lines are observed from the higher atomic number elements. As with the K series, a significant number of forbidden transitions and satellite lines are observed, but unlike the K series, where satellites arise almost exclusively from dual ionizations directly from primary photon impact, satellites in the L series arise mainly from auto-ionization.

2.10.1   The gold L spectrum

Gold has atomic number 79 and in the ground state has filled K, L, M, N, 5s, 5p and 5d levels plus one 6s electron. Figure 2-11 gives the transition diagram for

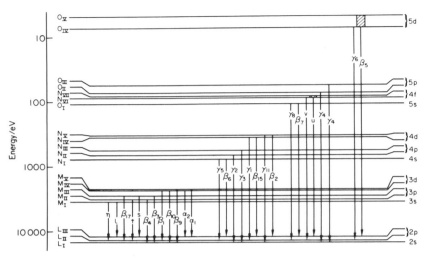

Fig. 2-11 Observed transitions in the gold L spectrum

gold and Fig. 2-12 shows the recorded L emission spectrum. Table 2-5 lists all of the normal and forbidden transitions along with their approximate relative intensities.

Comparison of the transition diagram and the emission spectrum shows the latter to consist of three major groups of lines, the α's the β's and the γ's. The α lines plus the majority of the β's come from transitions from the M sub-groups,

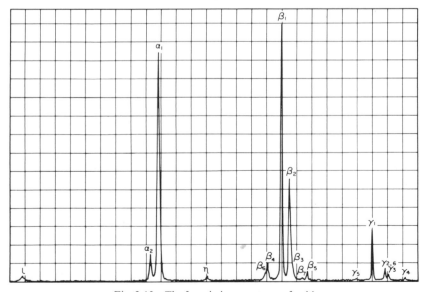

Fig. 2-12 The L emission spectrum of gold

TABLE 2-5.
The gold L spectrum

Line†	Wavelength	Transition	Approximate intensity
l	1.45964	$3s \to 2p^{3/2}$	2
(L)	1.41366	$3p^{1/2} \to 2p^{3/2}$	<1
(S)	1.35131	$3p^{3/2} \to 2p^{3/2}$	<1
α_2	1.28772	$3d^{3/2} \to 2p^{3/2}$	14
α_1	1.27640	$3d^{5/2} \to 2p^{3/2}$	100
η	1.20273	$2s \to 2p^{1/2}$	2
(β_{17})	1.12798	$3p^{3/2} \to 2p^{1/2}$	<1
β_6	1.11092	$4s \to 2p^{3/2}$	3
β_4	1.10651	$3p^{1/2} \to 2s$	8
β_1	1.08353	$3d^{3/2} \to 2p^{2/2}$	112
β_{15}	1.07188	$4d^{3/2} \to 2p^{3/2}$	<1
β_2	1.07022	$4d^{5/2} \to 2p^{3/2}$	45
β_3	1.06785	$3p^{3/2} \to 2s$	5
β_7	1.04974	$5s \to 2p^{3/2}$	1
(u)	1.04752	$4f \to 2p^{3/2}$	<1
β_5	1.04044	$5d \to 2p^{3/2}$	3
(β_{10})	1.02785	$3d^{3/2} \to 2s$	<1
(β_9)	1.02063	$3d^{5/2} \to 2s$	<1
γ_5	0.95559	$4s \to 2p^{1/2}$	1
γ_1	0.92650	$4d^{3/2} \to 2p^{1/2}$	22
γ_8	0.90989	$5s \to 2p^{1/2}$	<1
(V)	0.90837	$4f^{5/2} \to 2p^{1/2}$	<1
γ_2	0.90434	$4p^{1/2} \to 2s$ ⎫	4
γ_6	0.90297	$5d^{3/2} \to 2p^{1/2}$ ⎭	
γ_3	0.89773	$4p^{3/2} \to 2s$	3
(γ_{11})	0.88433	$4d^{5/2} \to 2s$	<1
γ'_4	0.86816	$5p^{1/2} \to 2s$	<1
γ_4	0.86703	$5p^{3/2} \to 2s$	<1

† "Forbidden" transitions are indicated by parentheses.

whereas the γ lines come mainly from N and O transitions. The most important exception to this generalization is the β_2 line which is associated with a transition from the N_V (4d) level. One immediately noticeable feature of the L spectra from the lower atomic number (about $Z = 40$) elements is the absence of this β_2 line. Other important β lines which are associated with transitions beyond the M levels are the β_6, β_7 and β_5.

Unlike the K spectra, the α's are no longer the dominant feature in terms of intensity. The β_1 is frequently as strong, if not stronger (as in this instance) than the α_1. This point will be discussed in detail in the section on line intensities (p. 33).

Eight "forbidden" transitions occur in the gold L spectrum and these lines are placed in parentheses in Table 2-5. These lines are extremely weak and have little analytical significance.

2.10.2   The strontium L spectrum

The gold spectrum just discussed is typical of most L spectra from elements of atomic numbers greater than about 42. Below this, the electron depletion of the upper levels cause significant changes in the emission spectra. Strontium gives an L spectrum that is typical of the lower atomic number elements. Strontium is atomic number 38 and in the ground state has filled K, L, M, 4s, 4p and 5s levels, Figures 2-13 and 2-14 show respectively the transition diagram and the L emission spectrum. The marked differences between the strontium (Fig. 2-14) and gold (Fig. 2-12) are very evident. Since strontium has no electrons in the 5d, 5p or 4d levels, the γ lines are almost non-existent and only those associated with the 4p level (γ_2 and γ_3) are at all visible. The γ_5 should also be visible but is masked by the β_3 line. Also missing are the β_2 and β_5 lines.

The absence of certain lines is by no means the only difference between the two spectra. Due to the absolute differences in the binding energies of the L_{II}, L_{III}, M_I and M_V levels, the energy gap $L_{II} - M_I$ (η) is bigger than that for $L_{III} - M_V$ (α_1) in the case of gold, but is smaller in the case of molybdenum. The effect is that wherease the η line occurs to the short wavelength side of the α_1 in the case of gold, it occurs on the long wavelength side in the case of strontium.

As in the case of the K spectra (and for the same reason), the α_1 and α_2 lines of the lower atomic number elements are not resolved.

The last important feature of the strontium L spectrum that should be noted is the appearance of line broadening to the high energy side of all of the lines.

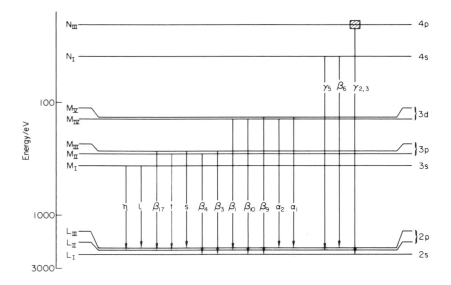

Fig. 2-13   Observed transitions in the strontium L spectrum

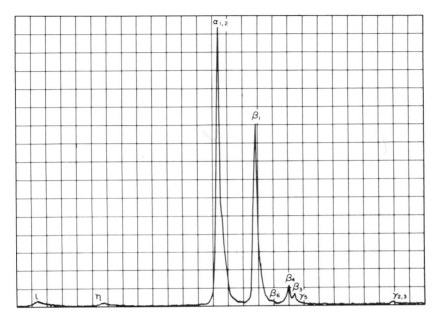

Fig. 2-14   The L emission spectrum of strontium

This is most noticeable in the cases of the $\alpha_{1,2}$ doublet and the β_1 lines, and will be seen to be very similar to that seen in the K spectra of the lower atomic number elements. Indeed, as in the case of the K spectra, the broadening is due to the occurrence of satellite lines, which in this instance are resolved neither from their parent lines nor from each other.

2.11   M SPECTRA

As might be expected from what has been described previously, M spectra are more complex and more variable than their K and L counterparts. In addition to the large number of levels that may be involved in transitions, it is also found that self-absorption causes significant changes in the emission spectrum.[18] Since the measurable wavelength region of most analytical X-ray spectrometers extends only to around 20 Å, the M spectra are less often encountered, and then only in the vacuum region (> 3 Å) see Fig. 2-15. Nevertheless, the strongest M lines are observable by the time the rare earth region is approached (i.e., $Z > 57$).

Figures 2-16 and 2-17 show respectively the transition diagram and M emission spectrum from tungsten. Tungsten is atomic number 76 and in the ground state has filled K, L, M, N, 5s, 5p and 6s levels with four electrons in the 5d level.

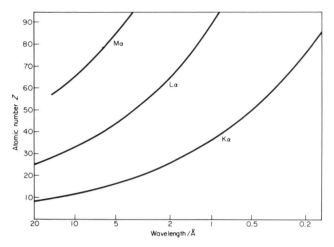

Fig. 2-15   Characteristic wavelengths as a function of atomic number

The majority of the M lines have not been assigned names, hence most of the lines indicated in Fig. 2-17 are identified by their transition states.

Some similarity will be seen between the tungsten M spectrum and the strontium L spectrum discussed previously. Both spectra are typified by a strong α and relatively strong β emission with weaker lines to the high energy side. Both spectra show line broadening to the high energy side of the α and β lines due in each case to the presence of auto-ionized satellites.

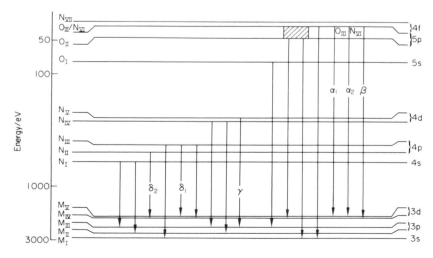

Fig. 2-16   Observed transitions in the tungsten M spectrum

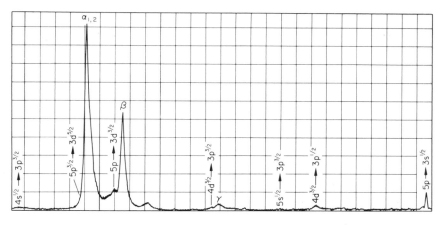

Fig. 2-17   The M emission spectrum of tungsten

2.12   RELATIVE LINE INTENSITIES

One immediately noticeable feature about the spectra shown in Figs. 2-4 to 2-7 is the considerable variation in the relative line intensities within a given K spectrum. Less obvious perhaps is the fact that the ratios of the intensities vary from element to element. Table 2-6 summarizes the data taken from the given spectra and shows that whereas the α_1 to α_2 intensity ratio is always constant and equal to 2 (if a high resolution spectrometer had been used, the same intensity ratio would have been found for β_1 and β_3) the intensity ratio of the sum of the $\alpha_1\alpha_2$ doublet to the sum of the $\beta_1\beta_3$ doublet is far from constant.

TABLE 2-6.
Characteristic line intensities for the K spectra

Element	$I(K\alpha_1)$	$I(K\alpha_2)$	α_1/α_2	$I(K\beta_{1,3})$	$\alpha_1\alpha_2/\beta_1\beta_3$
Sn	63	31.5	2	18	5.3
Cu	62	32	2	10	9.4
Ca	66	33	2	11	9.0
Al	—	72 —	—	—	36

Some light can be thrown into this if one considers that although both pairs of lines occur from the same type of transition, i.e., $p^{1/2} \rightarrow s$ and $p^{3/2} \rightarrow s$, the α's are involved with a Δn of 1 and the β_1 and β_3 with a Δn of 2. It is found that the probability of a transition occurring is related to an exponential function involving the difference in energy states of the electron in different orbitals. Figure 2-18 represents a qualitative attempt to illustrate what this really means. Here an atom is shown having filled orbitals up to and including the 4p level (e.g. Sn), thus all lines involving the transitions 2p → 1s, 3p → 1s and 4p → 1s are allowed. For simplicity, just the $K\alpha_1$, the $K\beta_1$ and $K\beta_2$ are shown. At the lower portion of the figure an indication is given of the way the electron energy varies as a function of the distance from the nucleus and the minima represent the levels K, L, M and N. Whereas a line arising from a Δn of 1, e.g. the $K\alpha_1$ line, represents a simple transference from one level to the next, a line involving a Δn of 2, e.g., the $K\beta_1$ line, involves transference across the L level. Similarly, a Δn of 3 giving the $K\beta_2$ line involves transference across both L and M levels. Simply from the consideration of electrostatic repulsion, a transference involving $\Delta n = 1$ is far more likely or "probable" than one involving $\Delta n = 2$. In practice, the situation is more complicated than this simple model since the levels themselves are split and may or may not all be filled. Further, the outer levels may involve bonding, and a molecular orbital approach rather than an atomic orbital approach is required.

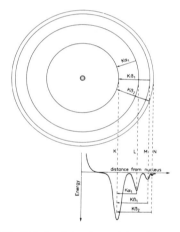

Fig. 2-18   Relative intensity and transition probability

Nevertheless, the concept of "transition probability" is an important one and for further discussion, the reader is recommended to consult works dealing with theoretical physics.[19,20]

Multiple transitions arising from a given set of states, e.g., the $K\alpha_1 K\alpha_2$ or the $K\beta_1 K\beta_3$ are called "multiplets". Essentially, a multiplet is a group of lines whose initial states arise from a single configuration. It is often found that the relative intensities of multiplet lines are predictable since they are directly related to the j quantum numbers. The Burger–Dorgelo sum rule[21] states that "for transitions between two term multiplets, the two sums of the intensities of all transitions from any one level are in the ratio of the statistical weights of these levels". The statistical weight W of a term characterized by the quantum number j is $2j + 1$, thus for the $K\alpha_1$ and $K\alpha_2$ lines, the intensity ratio should be

$$\frac{I(K\alpha_1)}{I(K\alpha_2)} = \frac{W(K\alpha_1)}{W(K\alpha_2)} = \frac{(2j + 1)K\alpha_1}{(2j + 1)K\alpha_2} = \frac{(2 \times 3/2) + 1}{(2 \times 1/2) + 1} = \frac{4}{2} \text{ or } \frac{2}{1}$$

Since the $K\beta_1 K\beta_3$ are also associated with j's of $3/2$ and $1/2$, respectively, they will also be in the ratio of $2:1$.

Obviously, this rule will tend to break down where unfilled or hybridized orbitals are involved, as for instance in the case of the oxygen K spectrum considered earlier. Nevertheless, the two maxima shown on the oxygen K spectrum in Fig. 2-10 are roughly in the proportion of $2:1$ illustrating the $2p^{3/2}$ and $2p^{1/2}$ character of the extremes of the 2p band.

Although L spectra are more complex both in structure and distribution of relative line intensities, the same basic concept apply as in the K series. Thus, for K series multiplets, the strongest lines were always associated with the largest value of $2j + 1$, and for the L series it will be seen that within groups the strongest lines again come from levels with the largest values of $2j + 1$. For instance, in transitions from M to L, the strongest lines are the $\alpha_1\alpha_2$ and β_1, all of which come from the upper (3d) levels. (As has been previously stated, forbidden transitions do not obey the normal rules, hence although the β_9 and β_{10} also come from an upper level, they are not necessarily expected to be strong lines). Similarly in N to L transitions, the β_2 and γ_1 lines come from the upper N level (4d) and again are the strongest of the group.

The relative intensities of $\alpha_1\alpha_2$ and β_1 can also be predicted[22] from the sum rule as were the $K\alpha_1$ and $K\alpha_2$, although in this instance it is necessary to ignore the small energy gaps between M_{IV} and M_V, and L_{II} and L_{III}. Thus

$$\frac{W\alpha_1 + W\alpha_2}{W\beta_1} = \frac{W_{L_{III}}}{W_{L_{II}}} = \frac{(2 \times 3/2) + 1}{(2 \times 1/2) + 1} = \frac{2}{1}$$

also

$$\frac{W\alpha_2 + W\beta_1}{W\alpha_1} = \frac{W_{M_{IV}}}{W_{M_V}} = \frac{(2 \times 3/2) + 1}{(2 \times 5/2) + 1} = \frac{2}{3}$$

assume $I\alpha_2 = 1$, then

$$1 + I\alpha_1 = 2I\beta_1$$

and

$$3[1 + I\beta_1] = 2I\alpha_1$$

from which $I\alpha_1 : I\alpha_2 : I\beta_1 = 9:1:5$.

The M series line intensities, as might be expected, are very difficult to predict but again the strongest line comes from the uppermost level within a group, provided of course, that these lines are not forbidden lines.

References

(1) Siegbahn, M. and Leide, A., *Phil. Mag.* **38**, 647 (1914).

(2) Bearden, J. A., *Rev. Mod. Phys.* **39**, 78 (1967).

(3) Bearden, J. A., *Phys. Rev.* **137**, 455 (1965).

(4) Taylor, B. N., Langenberg, D. N. and Parker, W. H., *Sci. Am.* 144 (1970).

(5) Bearden, J. A., U.S. Atomic Energy Commission Report NYO-10586, 1964.

(6) Burhop, E. S., *The Auger Effect* University Press, Cambridge, 1952.

(7) Campbell, W. J. and Gilfrich, J. V., *Anal. Chem.* **42**, 248R (1970).

(8) Fink, R. W. *et al.*, *Rev. Mod. Phys.* **38**, 513 (1966).

(9) Suortti, P., *J. Appl. Phys.* **42**, 5821 (1971).

(10) Hirsh, F. J. Jr., *Rev. Mod. Phys.* **14**, 45 (1942).

(11) Candlin, D. J., *Proc. Phys. Soc. A* **68**, 322 (1955).

(12) Aberg, T., *Phys. Rev.* **156**, 35 (1967).

(13) Siegbahn, K. *et al.*, *ESCA Atomic, Molecular and Solid State Structure Studied by means of Electron Spectroscopy* Almquist and Wiksells, Upsala, 1967.

(14) Fischer, D. W., *Advan. X-ray Anal.* **13**, 173 (1969).

(15) Fabian, D. J., *Soft X-ray Band Spectra*, Academic Press, London, 1968.

(16) Fischer, D. W., *Advan. X-ray Anal.* **13**, 159 (1969).

(17) Urch, D. S., *Quart. Rev. (London)* **25**, 343 (1971).

(18) Fischer, D. W. and Baun, W. L., *Advan. X-ray Anal.* **11**, 230 (1967).

(19) Garbuny, M., *Optical Physics*, Academic Press, New York, 1965, p. 87.

(20) Kuhn, H. G., *Atomic Spectra*, University Press, London, 1962.

(21) Dorgelo, H. B. and Burger, H. C., *Z. Physik* **23**, 258 (1924).

(22) Compton, A. H. and Alison, S. K., *X-rays in Theory and Experiment* Van Nostrand, New York, 1935, p. 649.

CHAPTER 3

physics of X-rays

Two major processes are involved in the interaction of X-rays with matter, these being scatter and photoelectric absorption. An X-ray beam is attenuated on passing through an absorber and the degree of attenuation will depend upon both scattering and absorption processes. When a beam of X-ray photons falling onto an absorber is scattered—mainly by the loosely bound outer electrons—and where no transference of energy is involved in the scattering process, the scattering is said to be coherent. When the scattered wavelength trains interfere with each other, diffraction phenomena may also occur. In X-ray spectrometry the diffraction process is of particular importance since this serves as the basis of the wavelength separation technique in the crystal spectrometer. Incoherent (Compton) scatter occurs where a small fraction of the energy of an incident X-ray photon is transferred to a loosely bound electron of the target element. In this instance, the wavelength of the incoherently scattered line is slightly longer than the incident wavelength.

Photoelectric absorption occurs when the electron of the target atom is completely removed (excited) from its initial site. In the analytical X-ray region, the major contributors to photoelectric absorption are the inner electrons of the absorber atoms, and since the binding energies of the inner electrons vary enormously with atomic number, characteristic X-radiation excited within a sample may be subject to severe absorption problems. This gives rise to so-called matrix effects, which greatly complicate the mathematical relationship between measured characteristic wavelength intensities and chemical composition.

The following sections discuss in detail the various absorption processes and their significance in the efficiency of the excitation of characteristic wavelengths. Interference and diffraction phenomena are also outlined along with a discussion of the crystalline state.

3.1 ABSORPTION OF X-RADIATION

When a beam of X-radiation falls onto an absorber, a number of different processes may occur. The more important of these are illustrated in Fig. 3-1.

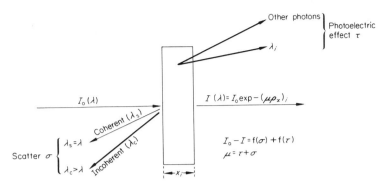

Fig. 3-1   Interaction of X-rays with matter

In this example a monochromatic beam of radiation of wavelength λ and intensity I_0 is incident on an absorber i of thickness x_i and density ρ_i. The fate of each individual incident X-ray photon is governed by the following three processes.

Absorption. A certain fraction (I/I_0) of the radiation may pass through the absorber. Where this happens, the wavelength of the transmitted beam is unchanged and the intensity of this transmitted beam $I(\lambda)$ is given by

$$I(\lambda) = I_0 \exp\left[-(\mu_i\rho_i x_i)\right] \qquad (3\text{-}1)$$

where μ_i is the mass absorption coefficient of absorber i for the wavelength λ, in units of g cm^{-2}.

The photoelectric effect. From the above it will be apparent that an amount of intensity equal to $(I_0 - I)$ has been lost and the majority of this is lost due to so-called photoelectric absorption (i.e., ejection of electrons due to photon impact). Photoelectric absorption (usually designated τ) will occur at each of the energy levels of the atom and the total photoelectric absorption. τ_{total} will be determined by the sum of each of the individual absorptions. Thus

$$\tau_{\text{total}} = \tau_K + [\tau_{L_I} + \tau_{L_{II}} + \tau_{L_{III}}] + [\tau_{M_I} + \tau_{M_{II}} + \tau_{M_{III}} + \tau_{M_{IV}} + \tau_{M_V}]\ldots + \tau_n$$

$$(3\text{-}2)$$

where τ_n represents the outermost level of the atom containing electrons. All radiation produced as a result of electron transitions following ejection of orbital electrons will have a wavelength longer than λ. Also, not all of the radiation produced will be X-radiation, hence in Fig. 3-1 the photoelectric effect is shown to give rise to X-radiation (λ_i) from the absorber and to other photons.

Scatter. Scattering can occur when an X-ray photon interacts with one of the electrons of the absorbing element. Where this collision is elastic i.e., no energy is lost in the collision process, the scatter is said to be coherent (Rayleigh scatter). Since no energy change is involved, the coherently scattered radiation

(λ_s) will retain the same wavelength as the incident beam. As will be shown later, X-ray diffraction is simply a special case of coherent scatter. It can also happen that the X-ray photon loses part of its energy in the collision process, especially where the electron is only loosely bound. In this case, the scatter is said to be incoherent (Compton scatter) and the wavelength (λ_c) of the incoherently scattered photons will be longer than λ. The total scatter σ is made up of both coherent and incoherent terms, thus:

$$\sigma = Zf^2 + (1 - f^2)$$
$$\text{coherent} \quad \text{incoherent}$$
$$(3\text{-}3)$$

where f is the so-called atomic structure factor.[1]

3.1.1   Mass absorption coefficient

The value of the mass absorption μ referred to in equation (3-1) is a function both of the photoelectric absorption and the scatter, in fact

$$\mu = \tau + \sigma \tag{3-4}$$

However, τ is generally large in comparison with σ and by combination of equations (3-2) and (3-4)

$$\mu \simeq \tau_K + [\tau_{L_I} + \tau_{L_{II}} + \tau_{L_{III}}] \ldots \tau_n \tag{3-5}$$

The way in which the mass absorption coefficient varies with wavelength is illustrated in Fig. 3-2. This shows the absorption curve for tungsten and the value of the mass absorption coefficient increases steadily with wavelength, in fact as the third power of the wavelength (and fourth power of atomic number of the absorber). It will also be seen that the curve has very sharp discontinuities, called absorption edges, indicated K, L_I, L_{II}, L_{III}, etc.

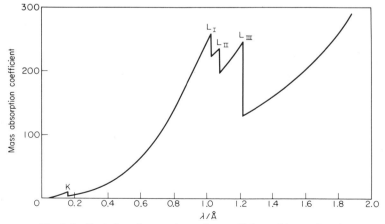

Fig. 3-2   Variation of mass absorption coefficient with wavelength

These absorption edges correspond to the binding energy of electrons in the appropriate levels. Where the absorbed wavelength is shorter than the wavelength of one of the edges an electron from the corresponding level can be excited. For instance in Fig. 3-2 an absorbed wavelength of 0.1 Å is shorter than the K absorption edge of tungsten (0.178 Å) and hence photoelectric absorption in the K level can occur. For an absorbed wavelength of 0.2 Å, however, photoelectric absorption in the K level certainly cannot occur. Thus, in general terms, each time the wavelength increases to a value in excess of a certain absorption edge one of the terms in equation (3-5) drops out, with a corresponding decrease in the value of the total absorption term. Mass absorption coefficients are well documented for most of the X-ray region and data are readily available in tabular form.[2]

The total absorption of a specimen is determined by summing all of the individual contributions for each element j in the matrix. Thus the total absorption is

$$\mu_{\text{total}} = \sum_j \mu_j w_j \tag{3-6}$$

where μ_i is the mass absorption coefficient of element i for the measured wavelength and w_i is the equivalent weight fraction. Note that whereas for concentration c we have $\sum_j c_j = 100\%$, for weight fraction w we have

$$\sum_j w_j = 1.$$

3.2 EXCITATION OF CHARACTERISTIC RADIATION

Figure 3.3 illustrates the practical situation encountered in the type of geometrical configuration employed in conventional X-ray spectrometers. Here a beam of radiation strikes the specimen at an (average) angle ψ_1 and a fraction of the excited (secondary) fluorescence beam leaves the specimen at an angle Ψ_2,

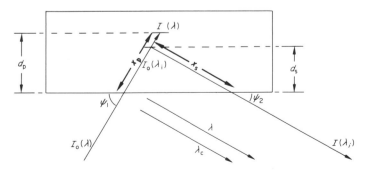

Fig. 3-3   Primary beam interaction in the X-ray spectrometer

defined by the primary collimator. The beam of primary radiation is of course polychromatic but for the sake of simplicity a single wavelength λ is considered in the sketch. The intensity of the primary wavelength is $I_0(\lambda)$ and a fraction of this beam $I(\lambda)$ travels a distance x_p reaching a depth d_p. The value of x_p will be determined by the mass absorption coefficient of specimen i for wavelength λ, i.e., $\mu_i(\lambda)$ and the density ρ_i of the specimen. Provided that the energy of the incident X-ray photons is greater than the excitation potential of i, characteristic lines from i will be excited. Again, just one of these wavelengths λ_i is considered in the sketch. Here the initial intensity of the secondary beam $I_0(\lambda_i)$ is considered at depth d_s, and the fraction of (measurable) secondary radiation $I(\lambda_i)$ will be dependent upon the take-off angle ψ_2, the mass absorption coefficient of the specimen for λ_i equal to $\mu_i(\lambda_i)$ and the specimen density.

In quantitative X-ray spectrometry it is important that the specimen be thick enough to exceed the critical depth of the secondary radiation. This critical depth occurs at the point where increasing d_s gives no significant increase in the measured fluorescence radiation. The critical depth can be calculated from the effective path length x_s and the take-off angle ψ_2,

$$d_s = x_s \sin \psi_2 \qquad (3\text{-}7)$$

Typical values for ψ_2 are 30–45° thus the critical depth is roughly one half of the effective path length of the fluorescence radiation.

The primary beam may also be coherently and incoherently scattered. Coherent scatter is the major cause of background in the measurement of secondary characteristic wavelengths. Incoherent scatter tends simply to "smear" the scattered continuum and to displace it to longer wavelengths. Any characteristic primary wavelengths, for example, from the anode material of the X-ray tube, will also be incoherently scattered and the scattering will appear as an additional broad peak to the long wavelength side of each coherently scattered line. Note from equation (3-3) that scatter is atomic number dependent, thus low average atomic number specimens will give higher backgrounds than high average atomic number specimens.

3.2.1 Primary and secondary absorption

We deduce from the foregoing that the intensity of secondary radiation will be dependent upon the absorption of the primary wavelength(s) by the specimen as well as by the attenuation of the secondary excited wavelength. A generalized scheme of the excitation process is given in Fig. 3-4 which illustrates the case of a certain element i with a wavelength λ_i being excited by continuous radiation from a primary X-ray source. The effective range of this continuous radiation lies between the short wavelength limit of the continuum λ_{min} and the absorption edge λ_{edge} of the excited element. The effectiveness of the excitation is to a first approximation given by the area of overlap of the absorption curve of element j and the continuum, indicated by $J(\lambda)$. Thus, in general, the most effective exciting

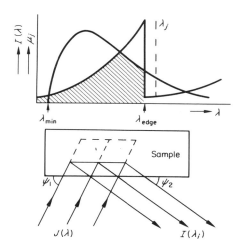

Fig. 3-4   Excitation of secondary radiation from the sample

radiation is that closest to the absorption edge and this is easily understood from the figure shown. The short wavelength primary radiation penetrates to a relatively great depth within the specimen, as its absorption by the sample matrix is usually less than that of the less energetic excited radiation. In addition, the design of most X-ray spectrometers is such that the average angle between the primary beam and the specimen, is much larger than the take-off angle for the secondary radiation (usually at least twice). This again leads to a greater depth of penetration of the primary beam.

One of the inevitable results of these two effects is that although the shorter wavelength region of the primary continuum may well lead to excitation, the characteristic radiation may be produced at such a depth that it cannot emerge from the specimen. Thus the most efficient exciting wavelengths are nearly always those closest to the absorption edge of the excited element, and of course these wavelengths must lie to the short wavelength side of the edge. The intensity of the measured wavelength λ_i is dependent upon the absorption prior to emergence from the specimen—this being the so-called secondary absorption effect. It will, however, also depend on the absorption by the specimen of the exciting continuum, this effect being called primary absorption. Since the excited characteristic line is monochromatic, it will be far more susceptible to variations in the absorption characteristics of the specimen than will be the polychromatic continuum. For this reason, the secondary absorption effect is invariably several times greater than that of the primary absorption. This in itself is a good thing from an experimental point of view, since secondary absorption effects are readily predictable and easily corrected for. Conversely, primary absorption effects are often difficult to predict and appropriate correction procedures are rather difficult to apply.

It is necessary that both primary and secondary absorption should be considered and in the section on quantitative analysis use will be made of the total specimen absorption $\sum_j \alpha_j w_j$ in which α_j is defined as

$$\alpha_j = [\mu_j(\lambda) + A\mu_j(\lambda_i)] \tag{3-8}$$

where $\mu_j(\lambda)$ is the mass absorption coefficient of a matrix element for a wavelength λ in the primary beam, $\mu_j(\lambda_i)$ is the mass absorption coefficient of the same matrix element j for the secondary wavelength λ_i. A is a geometric constant equal to $\sin \psi_1 / \sin \psi_2$.

3.2.2 Excitation efficiency

The quantity of characteristic radiation emitted from a specimen will be dependent upon the total available radiation from the source and the absorption characteristics of the atoms of the specimen for the primary and secondary X-ray photons. This quantity of characteristic radiation is generally expressed in terms of the excitation efficiency.

The total excitation efficiency for a given element i in a matrix made up of elements j is given by[3]

$$I(\lambda_i) = P_i \, w_i \int_{\lambda_{min}}^{\lambda_{edge}} J(\lambda) \frac{\mu_i(\lambda)}{\sum_j \alpha_j w_j} \tag{3-9}$$

where $I(\lambda_i)$ is the intensity of λ_i, w_i the weight fraction of i and P_i a proportionality constant, which is fixed for an equipment of fixed geometry operating under fixed excitation conditions. The integral of $J(\lambda)$ represents the tube spectrum between its excitation limits, i.e., the minimum wavelength of the continuum and the absorption edge of the element i.

3.2.3 Enhancement and third element effects

So far only direct excitation of the element i by the X-ray tube spectrum has been considered. In practice direct excitation is indeed the dominant excitation process but it is by no means the only way by which λ_i may be excited. Two other processes must also be considered. these being enhancement (sometimes called secondary fluorescence) and third element (sometimes called tertiary fluorescence) effects. The first of these effects occurs when a second (enhancing) element is first excited by the primary X-ray tube spectrum and then in turn excites element i. The second occurs when the primary tube spectrum excites an element (the "third" element) which in turn excites a second (enhancing) element, which, as before, excites element i. Figure 3-5 illustrates the three excitation processes. (a) shows the usual case of excitation of element i by the tube spectrum. (b) illustrates enhancement where the tube spectrum first excites λ_A from element A, followed by excitation of λ_i by λ_A. One condition for enhancement is that $\lambda_A < \lambda_i$ and it will be most pronounced when λ_A is of the same order

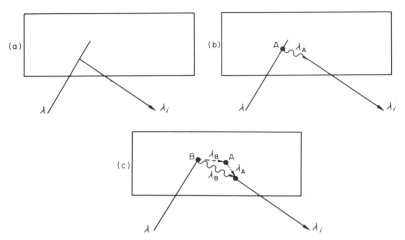

Fig. 3-5   Excitation by primary radiation, enhancement and third element effects

as λ_i. (c) illustrates the third element case where the tube spectrum first excites λ_B from a matrix element B. λ_B then excites λ_A from a second element A which in turn excites λ_i. In this instance, element B is the "third" element and A the enhancing element. A condition for the third element effect is that $\lambda_B < \lambda_A < \lambda_i$. It should be noted that in this instance an enhancement by λ_B on λ_i can also occur.

3.3   INTERFERENCE AND DIFFRACTION PHENOMENA

When light rays from different sources come into contact with matter, scatter and interference phenomena can occur. If, for example, a piece of blackboard chalk is crumbled and the chalk particles allowed to fall in bright sunlight to the ground, the falling particles are visible from a distance far greater than would be expected from their physical size. This is because the chalk particles scatter the light and the scattered light waves "interfere" with each other giving an enhanced light output in certain directions.

Figure 3-6 shows what happens when two or more trains of waves mutually interfere. Provided that the waves have a similar wavelength and provided that they are exactly in phase, the amplitude of the resulting wave is increased—an effect called "constructive" interference. Where the waves are exactly out of phase, they exactly cancel due to "destructive" interference. Such interference phenomena can occur when a single wave train is multiply scattered by an object and the scattered waves cross each other. This effect was demonstrated by Thomas Young in the early 1800's, and Fig. 3-7 illustrates his classic experiment. Here a light wave from a single narrow slit S_1 is allowed to fall onto two secondary slits S_2 and S_3 equidistant from the first. If either of the two secondary

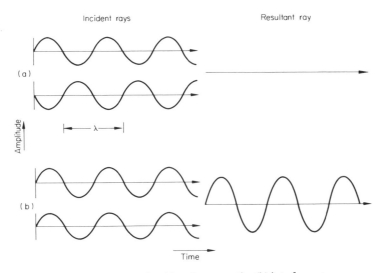

Fig. 3-6   Destructive (a) and constructive (b) interference

slits is covered, light passing through the uncovered slit gives a central bright band on the screen with some weaker bands to each side. The weaker bands are due to diffraction (or "bending") of the light beam as it passes through the slit. When both slits are open, each of the diffraction bands is crossed by evenly spaced dark vertical lines, arising from interference (interference fringes).

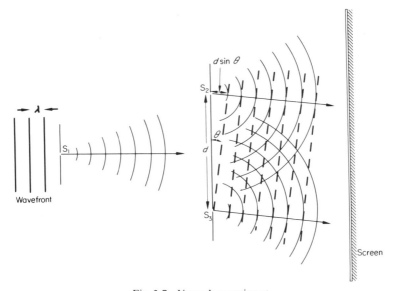

Fig. 3-7   Young's experiment

The interference fringes arise when the expanding wavelets from slits S_2 and S_3 merge together giving wavefronts, one set of which is indicated by the dashed lines in the figure. The position of the first fringe is given by the angle θ and it will be seen from the figure that

$$\sin \theta = \frac{\lambda}{d} \qquad (3\text{-}10)$$

where d is the distance between slits S_2 and S_3 and λ the wavelength of the light beam (i.e., the distance between the concentric rings). Other interference fringes will occur whenever other wavefronts can occur, this can be seen by drawing tangents to pairs of concentric rings, which are out of phase by an integral number greater than one. Thus in general

$$\sin \theta_n = \frac{n\lambda}{d} \qquad (3\text{-}11)$$

n is called the "order" of the line, thus for $n = 1$ the line is said to be first order, for $n = 2$ the line is second order, and so on.

Young's experiment illustrates that diffraction phenomena involve interference as well as diffraction. This is an important point and it will be seen later that whereas the *diffraction* of X-rays gives the means of measuring their wavelength, it is *interference* which leads to the enhancing effect giving a diffracted beam of sufficient intensity to measure.

3.3.1 Diffraction of X-rays

Equation (3-11) shows that in order to give a reasonable diffraction angle, λ and d need to be of about the same order of magnitude. For instance, if one tried to diffract X-rays with a wavelength of 1 Å (10^{-8} cm) with a slit of width 0.1 mm, the angle required to allow interference would be $\sin^{-1} 0.000001$. Thus it would seem that to diffract X-rays of wavelengths from 0.2 to 20 Å, over angular ranges θ of say 5 to 70°, d values of from 2 to 25 Å would be required. In 1912, Max von Laue[4] demonstrated that a crystalline material can be used as a three dimensional diffraction grating and, since the atomic distances in most crystalline substances are of the order of 1–10 Å, X-rays can be diffracted by use of a crystal. From this discovery have grown two major analytical techniques, X-ray diffractometry, in which X-rays of a fixed wavelength are diffracted by an ordered material, the diffraction angles thus giving information about interatomic distances, and X-ray spectroscopy, in which a crystal of known structure is used to separate a polychromatic beam of X-rays into individual wavelengths, the wavelength being calculable from the diffraction angle. In this instance, since the wavelength is characteristic of the atomic number, the intensity of the measured wavelength can be used to estimate the concentration of the respective element of which the line was characteristic.

3.3.2 The crystalline state

All substances are built up of individual atoms and nearly all substances have some degree of order, or periodicity, in the arrangement of these atoms. A crystal can be defined as a homogeneous, anisotropic body having the natural shape of a polyhedron. In practical terms whether a substance is crystalline or not can only be defined by the means that are available for measuring the crystallinity. In general, the shorter the wavelength the smaller the crystalline region that is able to be recognized. The visual extremes of periodicity are obvious: for example, a crystal of potassium chloride is as obviously crystalline as water is not. As far as an X-ray diffractometer is concerned fluids could not be defined as crystalline but probably 95 % of all solids could. Solids that have a very low degree of order are generally called glassy materials and these may frequently show some fluid properties. Although liquids may have some degree of order they cannot be defined as having crystalline properties hence the word crystalline is generally taken to be synonymous with the solid state. On the other hand, both glassy materials and liquids often give some diffracted intensity, generally in the form of one or two broad maxima at low diffraction angles.

In X-ray spectrometry we wish to ensure that the diffracting crystal is both efficient and specific in its ability to separate wavelengths. Thus, so-called "analysing crystals" are chosen for their high diffraction efficiency and only highly ordered arrangements are generally suitable.

Since a unique value of the interplanar spacing of the crystal is required, it is necessary to cleave the crystal in such a way that only the planes in the crystal that we want to diffract are allowed to do so. Since a crystal is built up of an ordered three dimensional array of atoms, it will contain many planes of high atomic density. For example, Fig. 3-8 shows a typical crystal lattice where the atoms are arranged in cubic symmetry and here several sets of planes of high atomic density are shown. It is necessary that one be able to define which sets of planes in the crystal are being considered and for this purpose a system of so-called "Miller indices" has been developed.

3.3.3 Miller indices

A point in space can be defined by reference to the basic axes x, y and z. For example, any point in a cube can be determined by a vector in space, this being the vectorial sum of the distances along the axes x, y and z, having the elementary values of a, b and c. Planes can be defined by their intercepts with the three basic axes x, y and z. A plane parallel to one of these axes has one intercept of infinity. To avoid the introduction of the concept of infinity and subsequent difficulties in calculation, it is normal to use the reciprocals of the intercepts. These reciprocals are first cleared of fractions and are then referred to as the "Miller indices" identified by h, k and l. Figure 3-8 indicates different sets of planes defined by their Miller indices. Note that the values of the Miller indices are always placed in parentheses.

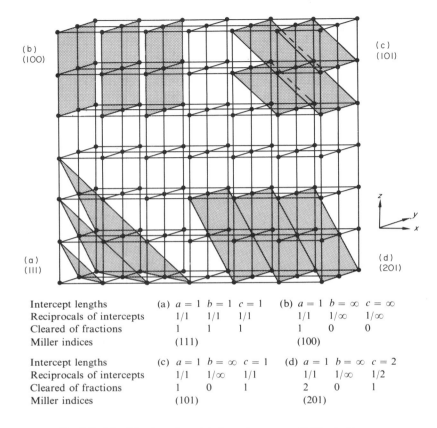

Intercept lengths	(a) $a = 1$ $b = 1$ $c = 1$	(b) $a = 1$ $b = \infty$ $c = \infty$
Reciprocals of intercepts	1/1 1/1 1/1	1/1 1/∞ 1/∞
Cleared of fractions	1 1 1	1 0 0
Miller indices	(111)	(100)
Intercept lengths	(c) $a = 1$ $b = \infty$ $c = 1$	(d) $a = 1$ $b = \infty$ $c = 2$
Reciprocals of intercepts	1/1 1/∞ 1/1	1/1 1/∞ 1/2
Cleared of fractions	1 0 1	2 0 1
Miller indices	(101)	(201)

Fig. 3-8   Identification of crystal planes by means of Miller indices

3.3.4 Conditions for the diffraction of X-rays

Figure 3-9 illustrates three parallel waves abc, a'b'c' and a"b"c" striking a set of crystal planes at an angle θ and being scattered. As in the case of the light rays previously discussed, reinforcement will occur when the difference in the path lengths is equal to a whole number of wavelengths. This path length difference is equal to d'b' + b'e' for ray a'b'c', and db" + b"e for ray a"b"c". These distances are equal to $2d \sin \theta$ and $2 \times 2d \sin \theta$ respectively; thus, the general rule for reinforcement is

$$n\lambda = 2d \sin \theta \qquad (3\text{-}12)$$

This condition was first formulated by Bragg and for this reason the expression given in equation (3-12) is called Bragg's law. The similarity between Bragg's law and the expression given in equation (3-11) will be immediately apparent.

The addition of further sets of planes to the model serves to give more constructive interferences and hence enhances the intensity of the diffracted beam.

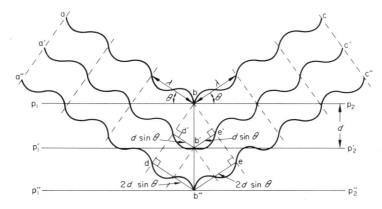

Fig. 3-9   Condition for the diffraction of X-rays

Again continuing the comparison with the light optical case previously discussed, where the path length differences between successive waves is equal to one wavelength one refers to a first order reflection. Where the path length differences is equal to two wavelengths, one refers to a second order reflection, and so on. Figure 3-10 illustrates the case for first, second and third order reflections, and here it will be seen that higher orders of a given wavelength will be diffracted at large values of θ by a given d spacing.

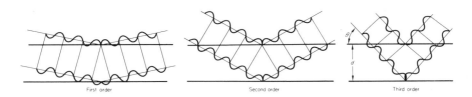

Fig. 3-10

3.3.5   Efficiency of X-ray diffraction

Since it is the electrons of the atom that coherently scatter the impinging wavelength and hence cause diffraction, the efficiency of a certain analysing crystal in diffracting X-radiation will be dependent upon the distribution of electrons within a given atom, as well as on the arrangement of all of the atoms in the lattice of the crystal.

An atom consists of a nucleus with many electrons moving around it in discrete orbits. When the angle between the impinging radiation and the direction of the observation is zero, there will be no phase difference between the waves and the resulting wave will have the maximum possible amplitude.

As this angle increases the waves of the outer electrons will gradually get out of phase. Their contributions then start to partially cancel and the amplitude of the resultant wave diminishes. The scattering power f of an atom is thus dependent upon its atomic number Z and the direction of the observation. The factor f is called the "atomic scattering factor" and is defined as the ratio of the amplitude of the wave scattered by an atom to the amplitude of the wave scattered by one electron. The atomic scattering factor will also be dependent upon the wavelength of the incident radiation since it is the magnitude of the path length difference relative to the wavelengths that is important (Fig. 3-11). In general, the form of the variation of f with the wavelength λ and the diffraction angle θ is that f decreases as $(\sin \theta)/\lambda$ increases. This is one of the reasons why higher order reflections (which occur at higher angles) are less intense than low order reflections.

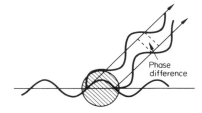

Fig. 3-11   Phase difference due to scattering from different parts of the atom

The total amplitude of the wave diffracted by a crystal lattice is formed by the vector summation over all elementary cells. The resultant wave of one elementary cell is made up of the individual contributions of the waves scattered by all atoms in the cell, each one with its correct amplitude f and its correct phase difference with respect to an arbitrarily chosen origin. This total scattering power of an elementary cell is called the structure factor F which can be expressed by the expression:

$$F(hkl) = \sum_{j=1}^{n} f_j(hkl) \exp \left[2\pi j(hx_j + ky_j + 1z_j) \right] \tag{3-13}$$

In this expression $F(hkl)$ and $f(hkl)$ refer to a given Bragg reflection, thus to a certain direction of observation. The factors x_j, y_j and z_j refer to the coordinates of the $j = 1 \ldots n$ atoms in the elementary cell. This sum may amount to zero for specific values of x_j, y_j and z_j for similar atoms having the same f-value. For example in the case of an elementary cubic cell having similar atoms at the corners and in the centre point at $\frac{1}{2} \frac{1}{2} \frac{1}{2}$, all reflections with hkl indices for which the condition $h + k + 1 = $ odd will show a value of $F = 0$. This fact is useful in that certain analysing crystals will not give odd order reflections and this helps to reduce experimental differences encountered in cases of harmonic overlap (see p. 88).

A perfect single crystal will give far less diffracted intensity than might be expected, the reduction in intensity being due mainly to secondary reflections of the diffracted beam from the undersides of the atomic planes. In an "ideally imperfect" crystal each separate crystal is made up of many small crystallites all slightly misorientated with respect to each other. In this instance the chance of secondary reflections is reduced (since a crystallite above the diffracting crystallite is not in the correct orientation for a secondary diffraction process). A slight displacement of the orientation of the crystallite is sufficient to bring the originally mis-orientated crystallites into the correct orientation. Diffraction lines thus show a significant breadth due to this mosaic effect as the individual crystallites will reflect a given "d" spacing over an angular range ($2\theta \pm \delta 2\theta$). Many instrumental factors, for example collimation, plus the natural wavelength distribution will contribute to the broadening of diffraction lines (see also p. 85).

In the section on the design of the spectrometer mention will be made of the need to match the spacing of the collimator with the mosaic of the crystal. In certain instances it is desirable to increase the mosaic structure of a "single" crystal by deliberately abraiding its surface.

References

(1) Compton, A. H. and Allison, S. K., *X-rays in theory and experiment*, Van Nostrand, New York, 1935, p. 140.

(2) Dewey, R. D., Mapes, R. S. and Reynolds, T. W., *Handbook of X-ray and microprobe data*, Progress Nuclear Energy, Series 9, Analytical Chemistry Vol. **9**, Pergamon, New York (1969).

(3) Jenkins, R. and de Vries, J. L., *Practical X-ray Spectrometry*, 2nd Edn., Macmillan, London, 1969, p. 22.

(4) Edward, P. P. (Ed.), *Fifty years of X-ray diffraction*, Oosthoek, Utrecht, 1962, p. 31.

CHAPTER 4
instrumentation

The instrumentation required to carry out X-ray spectroscopic measurements comprises three major portions—the primary source unit, the spectrometer itself and measuring electronics. The primary source unit is generally a sealed X-ray tube powered by a highly stabilized high voltage generator capable of delivering two to five kilowatts of power at up to one hundred thousand volts. The spectrometer may be either a wavelength dispersive or an energy dispersive system. The wavelength dispersive spectrometer relies on the diffracting property of a large single crystal to disperse the polychromatic beam of excited radiation. Flat, curved and logarithmically curved crystals have all been employed in this context—flat crystals being, in general, more suited to use in scanning spectrometers and curved crystals in fixed channel spectrometers. The energy dispersive spectrometer relies on the proportional property of a suitable detector to give a distribution of voltage pulses equivalent to the distribution of photon energies from the sample. A multi-channel analyser is then employed to separate individual energies.

The measuring electronics associated with the spectrometer generally include a stabilized high voltage supply for the detector(s), a timer/scaler combination which allows interdependent time and count measurements to be made and a rate meter circuit for integration of the digital signal from the detector, allowing a permanent record to be made of wavelength versus intensity. The following sections discuss in detail the various instrumental components for both wavelength and energy dispersive spectrometers.

4.1 EXCITATION SOURCES FOR X-RADIATION

In principle, almost any high energy particle can be used for the excitation of characteristic X-radiation. Although the majority of commercial X-ray spectrometers are fitted with primary X-ray sources consisting of a sealed X-ray tube and stabilized high voltage generator, electrons, protons, ions and γ-rays can be, and indeed have been, used.

Most of the very early work in X-ray emission analysis was performed using direct electron excitation using the specimen as the anode of a cold cathode tube.[1] Several problems arise in the use of this type of system of which the most important are the heating effect of the electron beam, the need to keep the specimen under very high vacuum and the difficulty in maintaining a stable output from the tube. In the cold cathode tube the source of electrons is residual gas left in the tube; thus the stability, tube current and voltage are all very interdependent. The practical operation of these tubes was rather difficult, but fortunately the advent in 1913 of Coolidge's hot cathode X-ray tube[2] greatly reduced these difficulties. In the Coolidge tube the source of electrons is a tungsten filament and the tube current and voltage can be made almost independent parameters. In fact, the Coolidge tube did not immediately provide a solution to all of the problems outlined above because at that time nothing was known of the principle of secondary excitation, i.e., the use of primary X-rays to produce secondary (fluorescent) X-rays. This idea came much later[3,4] when Von Hevesey was looking for a direct method for the determination of hafnium in zirconium oxide and was faced with problems of differences in volatility between these two elements. Even the secondary excitation sources used at that time bore little similarity to their present day counterparts. Figure 4-1 shows a schematic diagram of one of the original Coster–Druyvesteyn secondary fluorescence tubes.[5] In this tube the specimen to be analysed was placed on an aluminium foil, opposite the anode of the tube. The chance of backscattered primary electrons impinging on the specimen was reduced by use of a focusing cylinder which was held at the same potential as the cathode.

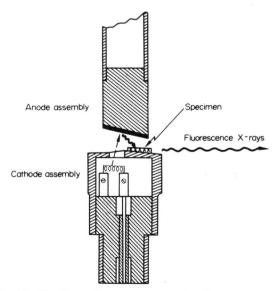

Fig. 4-1   The Coster–Druyvesteyn secondary fluorescence tube

For half a century the Coolidge tube, Fig. 4-2, has remained the time honoured source in all branches of X-ray analytical and radiographic work. Its usefulness being due mainly to its inherent high stability and the ability to utilize it as a completely independent source, thus allowing the sample to be left in air or, at the worst, in a relatively poor vacuum. The Coolidge tube has its own limitations and in the specific case of X-ray fluorescence spectrometry one of the major problems is the relatively low output of X-radiation with wavelengths longer than 10 Å or so. Until recent years, this has not posed too much of a problem since little analytical X-ray spectrometry has been attempted in the >10 Å region, which in terms of atomic numbers means elements below magnesium ($Z = 12$). This is not, however, because the spectroscopist is not interested in the $Z = 1$ to $Z = 11$ range, but because these elements (H, He, Li, Be, B, C, N, O, F, Ne, Na) are extremely difficult to analyse. In this region, the problem is really one of providing an entirely different type of equipment configuration, i.e., very large "$2d$" value crystals, detectors with windows of low absorption, the means of handling band spectra rather than line spectra and so on.

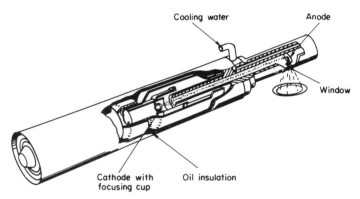

Fig. 4-2   A modern X-ray tube

Recent exploitation of the vacuum ultraviolet soft X-ray region for studies in space spectroscopy, plasma spectroscopy, and use of variations in emission wavelengths and intensities for the study of chemical bonding have tended to bring the appreciated, but often unconsidered, limitations of the Coolidge tube to the fore. Nevertheless, for all its limitations, the majority of workers in this field consider that the sealed, hot cathode tube remains by far the best alternative for the conventional X-ray analytical region.

4.1.1   Requirements of an X-ray source

There are four basic requirements of an X-ray source,
(a) Sufficient photon output over the required wavelength range.

(b) High stability (generally better than 0.1 %).

(c) Ability to work at reasonably high potentials (i.e., to around 60 to 100 kV.

(d) Freedom from too many interfering characteristic X-ray lines.

It is best to consider tube output in terms of the product of X-ray counts per second per watt and tube input power in watts. In general, a standard X-ray tube has good stability and a wide voltage range. Also the maximum power is reasonably high (values normally between 2 and 5 kW), but for long wavelengths the counts per second per watt are often barely sufficient.

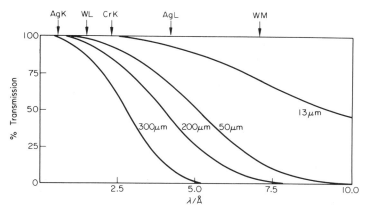

Fig. 4-3   Transmission characteristics of beryllium X-ray tube windows

This is essentially a problem of window absorption, and its magnitude is illustrated in Fig. 4-3, which shows window transmission as a function of wavelength for different thicknesses of X-ray tube windows. From the curves shown it is evident that in order to obtain a high output of long wavelength radiation it is vital for the X-ray tube window to be as thin as possible. Unfortunately, there is a great problem that must be solved before ultra-thin windows can be used and this is one of the window heating caused by electrons backscattered from the anode of the tube. The effect is illustrated in Fig. 4-4.

Two parameters are of importance in this phenomenon, these being the backscatter function S and the angle α between the anode and the cone of backscatter. As the atomic number of the anode decreases, S also decreases, but α increases. Hence a much thinner window can be employed with, for example, a chromium anode than with a tungsten anode, not only because the electron backscatter is less, but also because less of these electrons hit the window. It is in fact possible to utilize a relatively thin window in a chromium anode tube, perhaps several hundred microns thick, whereas in the case of tungsten a window thickness of about a thousand microns is required. A severe limitation exists with the sealed X-ray tube, in that if one reduces the window thickness, large temperature gradients are formed across the window, causing rapid failure. Alternatively,

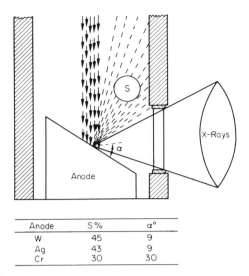

Anode	S %	α°
W	45	9
Ag	43	9
Cr	30	30

Fig. 4-4   Window heating due to electron back-scatter

if one tries to increase the total power, more backscattered electrons are produced, again giving the same window problem. Thus a three kilowatt tube of a given design and anode material will generally require a thicker window than a one kilowatt tube of similar design. Hence for light element work, too much store should not be placed on the rating of a given tube, but rather on what it will give in terms of fluorescent radiation from a given low atomic number element.

Many of the problems of window heating can be avoided by use of a "reverse potential" X-ray tube, in which all components, except the anode, are kept at ground potential.[6] The anode configurations of two types of reverse potential tubes are shown in Fig. 4-5. In each instance, since the body of the tube, and

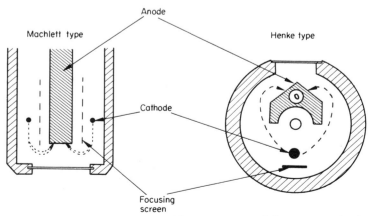

Fig. 4-5   Reverse potential X-ray tubes. All parts except anode kept at ground potential.

hence the tube window, is held at the same potential as the cathode, the electrons will be deflected away from the window, thus obviating the major cause of heating. All reverse potential tubes suffer from a problem of the cooling water being in direct contact with the high voltage on the anode. This water should be de-ionized to minimize power leakage.

4.1.2   Other configurations of X-ray tube

It is interesting to consider what other types of source are possible, which might retain the stability and the convenience of the sealed X-ray tube, but perhaps offer the possibility of greater output.

Figure 4-6 shows several other tube configurations which have been investigated and which employ magnets or grids[7,8] to reduce the electron bombardment on the window. No. 1 is the standard X-ray tube, electrons are indicated by wavy lines and X-rays by dashed lines. No. 2 is a tube incorporating a grid which is held at the same potential as the cathode, the function of the grid being to deflect the electrons from the area of the tube window. No. 3 utilizes a magnet around the window, as does No. 4, which is a windowless design[9,10] offering the use of X-rays or X-rays plus back-scattered electrons. In each case the function of the magnet is to deflect the electrons, in the first instance from the window and in the second to prevent the electrons striking the specimen, in cases where this is undesirable.

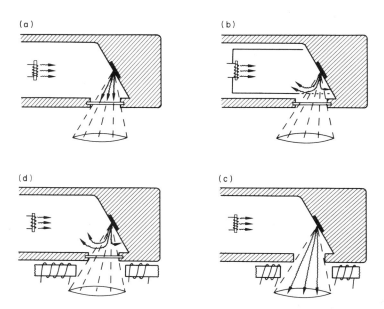

Fig. 4-6   X-ray tube configuration for long wavelength spectroscopy

What is important to consider is what gain is potentially available from any of the excitation systems mentioned. Table 4-1 summarizes data from different anode materials, with various window thicknesses.[11] First a 203 μm window (which can be considered to approximate to a standard sealed tube). Second a 13 μm window (which again approximates to the system employing reversed potential, grids or magnets). Finally, the windowless system (which approximates to the windowless X-ray tube and almost to electron excitation). Study of the data indicate a maximum gain of about 3 for the thin window system against a factor of 20 for the open window or direct electron excitation systems.

TABLE 4-1.
Anode excitation efficiencies and window effects for AlKα radiation expressed as counts $s^{-1}W^{-1}$

Anode	Window	Continuous	Characteristic	Back-scattered electrons	Total
W	none	57	70	884	1037
	none + screen	57	70	—	153
	13 μm	83	63	—	146
	203 μm	26	2	—	24
Ag	none	33	65	664	762
	none + screen	33	65	—	98
	13 μm	49	75	—	124
	203 μm	17	32	—	49
Cr	none	13	32	238	283
	none + screen	13	32	—	45
	13 μm	21	39	—	60
	203 μm	8	36	—	44

These figures must, of course, be related to the total power loading for the tube, which is generally several times higher for the sealed X-ray tube, than for open window configurations or for direct electron excitation.

4.1.3 Direct electron excitation

As was previously mentioned, much of the early work was carried out using direct electron excitation, and it is useful to consider the relative merits of the direct electron and X-ray tube excitation systems.[12] Generally speaking, the stability of the direct electron excitation method is somewhat inferior to that of the X-ray tube but against this the actual counts/s per W for a given element will always be higher with an electron source than with an X-ray tube operating at the same voltage and current. Conversely, the total usable power with direct electron excitation is somewhat less than that of the sealed X-ray tube, due mainly to the heating problem. Whereas in the sealed tube most of the heat is conducted away via the water cooling of the anode assembly structure, in direct electron excitation all of the heat must be dissipated via the specimen. This in turn presents certain problems in the preparation of specimens of non-conducting powders, but these can generally be overcome by carefully mixing and pelletizing the sample with graphite. The direct electron excitation system can be considered to offer a good alternative to the sealed X-ray tube particularly for the excitation of wavelengths longer than 5 Å.

4.1.4   Proton excitation

It was recognized as long ago as 1912 that X-rays were produced by proton bombardment[13] although not until the advent of particle accelerators in the 1930's were suitable sources generally available. In theory at least, the proton source offers several advantages over photon or electron excitation, one of the more important being that since the exciting protons are not deflected or decelerated by the sample, no continuous spectrum is generated.[14] Also the production efficiency of X-ray quanta is greater for wavelengths arising from transitions corresponding to outer shells of the atom.[15] Although the cost of the basic instrumentation is high, it has been estimated that there are already about 700 linear accelerators available in the world which could be used for this type of work. Current interest in the long term effects of trace elements in the environment is likely to further enhance interest in proton excitation methods, since it would appear that this technique is eminently suited to trace analysis. At the present state of the art, proton excited X-ray emission is potentially almost an order of magnitude more sensitive than neutron activation analysis and determinations down to 10^{-12} g have been performed.[16] Among the practical limitations encountered in trace analysis are contamination from the specimen backing material[17] and decomposition of the specimen if too high a radiation flux is employed.

4.1.5   Excitation by radioisotopes

The radioisotope offers a cheap, stable and compact source of radiation that is well suited to the excitation of a relatively short wavelength band.

The great disadvantage of the radioisotope source is that its total photon yield is several orders of magnitude lower than that of the sealed X-ray tube. This in turn precludes the use of conventional dispersive spectrometers based on analysing crystals and collimators. Thus analysing methods other than those based on diffraction phenomena have to be employed. Of these, the use of energy dispersion in the form of pulse height selection is the most versatile, but until recently, available detector systems, namely those based upon the scintillation and gas proportional counters, have had resolutions far inferior to the crystal spectrometer, particularly for the longer wavelengths.

This in turn means that multichannel pulse height analysers are not sufficient on their own and this limitation has led to the development of specialized techniques to aid energy isolation. These include the use of filters and specially fabricated sources which give preferential excitation of the required wavelength. Typical of these are the so-called γ-X sources.[18–20]

The gamma-source excites K X-radiation from the target and it is this X-radiation that is used for excitation of the required element(s) in the sample. It is now possible to fabricate γ-X sources giving up to 90 % of the energy output as target K radiation. γ-X sources are particularly useful as the basis of cheap, simple single element systems and have been especially applicable in the field of on-stream analysis. The advent of the high resolution Si(Li) energy dispersion spectrometer (see p. 92) has allowed, and indeed required, the use of excitation sources giving broad energy range excitation. Table 4-2 lists typical radioisotopes which have been used for this purpose, but current trends in technology are causing even these sources to be replaced by the miniature, high-powered X-ray tube.

TABLE 4-2.
Radioisotope sources used in energy dispersive spectrometry

Nuclide	Energy of radiation/keV	Half-life
Fe^{55}	5.9	2.7 years
Cl^{109}	19,87.8	470 days
Am^{241}	14–59.6	458 years
Co^{57}	122–137	270 days
Eu^{155}	71–77	1.8 years

4.2 DETECTION OF X-RAYS

The function of the X-ray detector is to convert the energies of the X-ray photons into voltage pulses which can then be counted, giving in turn a measurement of the total X-ray flux. Detectors used in modern X-ray spectrometers are almost invariably "proportional" detectors, and these have the property that the energy of an X-ray photon entering the detector determines the size of the output voltage pulse. This feature allows a form of voltage discrimination to be employed, called pulse height selection, which allows a selectable narrow band of voltage pulses to be passed on to the scaling circuitry giving, in effect, discrimination against unwanted photons.

Three types of detector are commonly employed, the gas flow proportional detector, the scintillation detector, and the semiconductor radiation detector. The first two of these are used mainly in crystal spectrometers where the spectrometer itself is the major means of dispersing the polychromatic beam of radiation from the specimen. Advantage is taken of the proportional characteristics of the detectors by employing pulse height selection to remove unwanted radiation, mainly in the form of harmonic reflections. The semi-conductor radiation detector has a high inherent energy resolution and this is commonly used in energy dispersive spectrometers, where the *only* form of dispersion is the detector itself. Both lithium-drifted silicon and lithium-drifted germanium have been used for this purpose. Gas flow proportional and scintillation counters have also occasionally been used in energy dispersive spectrometers but their energy resolution is rather poor and these systems are limited to certain element combinations, or to absorption dispersive systems where single or multiple filters are used to aid the selection of a certain wavelength region.

4.2.1 The gas flow proportional detector

When an X-ray photon interacts with an inert gas atom an outer electron may be removed leaving a positive ion. For example, in the case of argon:

$$Ar \xrightarrow{hv} Ar^+ + e$$

The resulting combination of electron plus positive ion is called an ion pair. The potential required to remove an outer electron from an inert gas is relatively small and even after allowing for the statistical factors governing the collision process the average energy (ε) required to produce an ion pair is still less than 30 eV. The actual value depends upon the atomic number of the gas and typical values are listed in Table 4-3. The number n of ion pairs produced by X-ray ionization is proportional to the energy E_0 of the radiation and is given by:

$$n = \frac{E_0}{\varepsilon} \tag{4-1}$$

TABLE 4-3.
Ionization values for the inert gases

Element	Density/g/dm^3	Atomic number	First ionization potential/ eV	Average energy to produce an ion pair/eV
He	0.179	2	24.6	27.8
Ne	0.900	10	21.6	27.4
Ar	1.874	18	15.8	26.4
Kr	3.708	36	14.0	22.8
Xe	5.841	54	12.1	20.8

In its simplest form the gas flow proportional detector consists of a hollow metal cylinder of about 3 cm diameter, carrying a thin (25–100 μm) wire along the radial axis, Fig. 4-7. This wire forms the anode of the counter and carries a

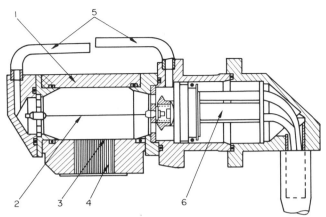

Fig. 4-7   The gas flow proportional counter:
1. Cylindrical Chamber
2. Anode Wire
3. Flow Counter Window
4. Collimator
5. Flow Gas Connections
6. Amplifier

potential of about one and a half thousand volts. The cylindrical casing is earthed and is filled at atmospheric pressure with a suitable gas mixture consisting of ionizable gas and quench gas, usually a mixture of 90% argon and 10% methane. It is important that this gas be free of electronegative impurities such as oxygen and carbon monoxide. A thin, low absorbing (1–6 μm) plastic window is fitted in the wall of the casing and is supported against external vacuum in the spectrometer by means of a simple collimator or a high transmission grid.

4.2.1(i)   Gas amplification

An X-ray photon entering the detector produces n ion pairs in the manner described. Provided that the potential across the detector is large enough to prevent recombination, the electrons move towards the anode and the positive ions to the casing of the detector body. Each primary electron does not travel very far before it strikes another argon atom. The electron may gain kinetic energy due to the accelerating field of the anode and, if sufficient energy is gained before it strikes a second argon atom, part of this energy may be given up in ionizing the second argon atom. Thus the initial number of electrons doubles and may double again and again, as the high field region close to the anode wire is approached. This effect results in a considerable "gain" or "multiplication" in the number of electrons N which finally reach the anode. The value of the gas gain G is given by

$$G = \frac{N}{n} = \frac{N\varepsilon}{E_0} \tag{4-2}$$

where G is typically in the range 10^4 to 10^5. From the point of view of the application of pulse height selection[21] it is vital that the proportional relationship between G and E_0 be maintained (hence the name proportional detector). It should be appreciated that this relationship no longer holds at high anode potential (i.e. high gas gains) where the detector approaches the Geiger region. Figure 4-8 illustrates the variation of the field with distance from the anode calculated for a 3 cm diameter flow counter fitted with a 25 μm anode wire. Curves are shown for different values of applied potential. A starting potential (E_r) of about 2×10^4 V/cm are required to start the amplification process, thus a cylindrical volume of amplification is formed concentric with the anode wire.

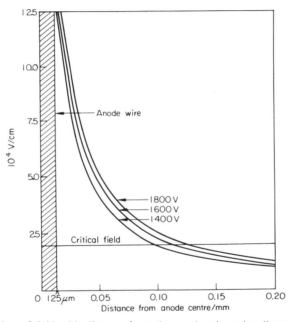

Fig. 4-8   Variation of field with distance from the anode wire, wire diameter 25 μm counter diameter 3 cm

For example, at 1400 V all of the amplification takes place within 0.10 mm of the anode (i.e., four wire diameters). In general the radius (I_r) of the volume of amplification is given by:

$$I_r = \frac{V_a}{E_r} \frac{1}{\ln (r_2/r_1)} \tag{4-3}$$

where V_a is the applied potential, r_2 the radius of the flow counter and r_1 the radius of the anode wire. Increasing the anode potential increases the volume of amplification, but at the same time the number of secondary ion pairs produced increases at a far greater rate. Thus the ion density increases considerably with applied potential so decreasing in turn the mean free path of the electron. The gas gain is thus a rather complex function dependent upon many factors. One useful form of the equation[22] is:

$$G = \exp\left\{2\left[\frac{V_a a r_1}{\ln(r_2/r_1)}\right]^{1/2}\left[\frac{V_a L_e}{V_i r_1 \ln(r_2/r_1)} - 1\right]\right\} \qquad (4\text{-}4)$$

where a is the differential ionization coefficient of the gas and V_i its effective ionization potential. L_e is the mean free path of the electron. It will be seen that the gas gain is an exponential function of the anode potential and so where stable pulse outputs are required (e.g. in pulse height selection), good stability of the applied voltage is of vital importance. Stabilities of the order of 0.05% are usually achieved.

4.2.1(ii) Pulse formation process

Figure 4-9 shows the basic circuitry of the proportional counter. A high voltage of 1.5 to 2.0 kV is applied via the leak resistor R_L. The functions of the leak resistor are to inhibit the leakage of charge on C_L during the decay period of the detector and to help quench the discharge process. The capacitance of the detector itself (indicated schematically by C_d) is typically of the order of 100 pF.

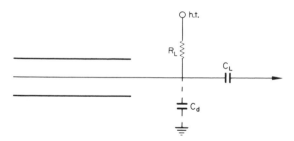

Fig. 4-9 Basic circuitry of the gas proportional counter

The size of the voltage pulse produced at C_L can be calculated quite simply; for example, the energy of a Cu Kα photon is 8.04 keV and from equation (4-1) it will be seen that about 300 primary ion pairs per photon (event) will be produced. Assuming a gas gain of 10^4 a total of 3×10^6 electrons per event will be produced. Since the charge carried by a single electron is 1.6×10^{-19} C the total charge per event will be 4.8×10^{-13} C. With a capacitance C_d of 10^{-10} F this would correspond to a voltage of

$$\frac{4.8 \times 10^{-13}}{10^{-10}} = 4.8 \text{ mV}$$

Since the sizes of voltage pulses produced are only of the order of millivolts some degree of amplification is required before the pulses are passed to the scaling circuitry (see p. 91). Figure 4-10 shows a schematic representation of the actual pulse formation process.

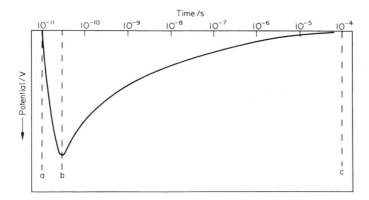

Fig. 4-10   Pulse formation process

Following the entry of the X-ray photon into the counter a series of ion pair avalanches is produced, the elapsed time between photon entry and avalanche production being something like 3×10^{-7} s. The electrons then separate and move to the anode, giving a sharp voltage drop (a → b). The positive ions then start to move towards the earthed casing, first rapidly then more slowly, due to the variation in the field between the anode wire and the earthed casing (b → c). This latter stage is rather slow, requiring something like 10^{-4} s, thus the pulses are normally clipped using an RC circuit with a time constant of about 1 μs.

4.2.1(iii)   Dead time

The relatively slow dissipation of the positive ion sheath has another effect upon the functioning of the counter in that as long as the ions are in the immediate vicinity of the anode, the field is lowered, thus preventing further avalanches. This gives rise to the so-called dead time of the counter. The pulse shaping circuitry also contributes significantly to this dead time effect and the average dead time value t_d including counter and associated circuitry is of the order of 1–2 μs in every second, and so a significant reduction in the count rate can occur. The measured count rate I_M will, therefore, always be lower than the true count rate I_T by an amount calculable from the usual dead time equation:

$$I_T = \frac{I_M}{1 - I_M t_d} \qquad (4\text{-}5)$$

For example, it will be seen that for a dead time of 2 μs the dead time loss at 10^5 counts/s is 20%. In fact equation (4-5) is only an approximated expression since the divisor on the right hand side of the equation should really be in the form of a binomial expansion thus:

$$I_T = \frac{I_M}{1 - I_M t_d + (I_M)^2 t_d^2/2! - (I_M)^3 t_d^3/3! \ldots} \qquad (4\text{-}6)$$

However, in equation (4-6) the error introduced by ignoring all but the first term of the expansion is just less than 2% at 10^5 counts/s and this error is normally acceptable. It may be as well to point out, however, that with the advent of higher powered X-ray tubes and faster scalers, count rates in excess of 10^5 are likely to become rather common, at which time the use of a more accurate expression for dead time than that given in equation (4-5) will be required.

4.2.1(iv) Counter resolution

One very important property of a counter is its so called resolution. Since there is a statistical process involved in the number of ionizing collisions made as the electrons move towards the anode[23] the pulse amplitude from the detector does not occur at a discrete voltage level but rather as a spread of voltages taking the shape of a distribution which is approximately Gaussian[24,25] (Fig. 4-11). The resolution (Res) of the counter is expressed in terms of the peak width at half height W expressed as a percentage of the maximum of the pulse amplitude distribution V, i.e.

$$\text{Res} = \frac{W}{V} \times 100 \qquad (4\text{-}7)$$

The usual statistical procedures can be applied to calculate the resolution of a counter from the number of primary electrons. Thus the variance should be proportional to the number n of primary electrons and the standard deviation proportional to $n^{1/2}$. Fano,[24] however, showed that the variance is only a fraction F of what would be expected. F, the Fano factor, usually takes a value between $1/2$ and $1/3$. Since for an argon counter $\varepsilon = 26.4$ the theoretical resolution of the counter for a photon of energy E keV would be

$$\text{Res} = F n^{1/2} \quad \text{or} \quad \frac{F \times 100}{n^{1/2}}\%$$

$$= 2.36 \times 100 \times \frac{(0.0264)^{1/2}}{E^{1/2}} = \frac{38.3}{E^{1/2}}\%$$

In practice it is found that the resolution equation is of the form

$$\text{Res} = \frac{KT}{E^{1/2}} \qquad (4\text{-}8)$$

where T is the theoretical factor (e.g. 38.3 for Ar) and K a factor which varies with the design of the counter, the diameter of the anode wire and, most importantly, on the cleanliness of the anode wire. Small particles of dirt which lodge on the anode cause local variations in the field with subsequent loss of resolution. In practice it is important to keep the value of K well below 1.2 since values in excess of this not only have an adverse effect on the successful application of pulse height selection but, equally important, can lead to significant shifts in the sizes of the pulses leaving the detector.[26] It is now becoming common practice for instrument manufacturers to pass the flow gas through a series of millipore filters before it enters the counter body.

4.2.1(v) Escape peak phenomena

When the energy of the X-ray photon entering the detector is greater than the absorption edge energy of the counter gas, characteristic radiation will be produced from the gas giving rise to an escape peak (see Fig. 4-11). The pulse amplitude distribution thus contains not only the pulse amplitude V_p from the discrete photon energy $E(\lambda)$ incident upon the counter, but an additional pulse V_e also occurs with an amplitude corresponding to the difference between the energy of incident radiation and the energy that has been lost, i.e., the energy of Ar Kα in the case where the counter gas contained argon as the ionizable gas. Thus:

$$V_p \propto E(\lambda)$$

$$V_e \propto E(\lambda) - E(\text{Ar K}\alpha)$$

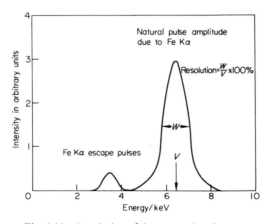

Fig. 4-11   Resolution of the proportional counter

When pulse height selection is employed for the removal of harmonic overlap in X-ray spectroscopy the situation can arise where although the photo pulse amplitude corresponding to an interfering element can be removed, the escape pulse amplitude still interferes with the pulses from the required element. This can be a serious problem in analytical X-ray spectrometry.[27] Figure 4-12 illustrates a classic case of escape peak interference obtained when phosphorus is measured in the presence of calcium. Line overlap occurs between P Kα and Ca2 Kβ and although the photo pulses from Ca Kβ can be removed, the escape pulses still interfere.

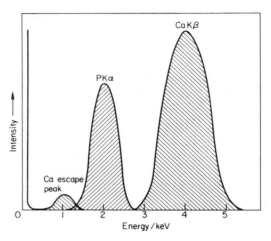

Fig. 4-12   Escape peak interference of calcium on phosphorus

4.2.2 The scintillation detector

The scintillation detector consists of two essential parts, the phosphor and the photomultiplier. The phosphor is typically a sodium iodide crystal doped with thallium and this converts the energy of the X-ray photon into light pulses of about 3 eV energy. The photomultiplier converts the light photons into voltage pulses which are in turn collected and amplified, as in the case of the gas flow proportional detector. Since solids are about 1000 times more dense than gases, the active volume of the phosphor is relatively small in comparison with the body of the gas flow proportional detector. The energy required to produce ion pairs in solids is almost an order of magnitude smaller than in gases, so in principle at least, the scintillation detector should give a greater number of ion pairs per incident X-ray photon with a corresponding smaller statistical fluctuation. Unfortunately, however, in practice there are inefficient processes involved in the conversion of free electrons into light pulses in the phosphor (about 20 % efficient), and in the conversion of light quanta into electrons in the photomultiplier (about 5 % efficient). These two combined factors make the total loss in efficiency about 100 and this makes the effective energy to produce an electron about 300 eV.

In comparison with the gas flow proportional detector, the energy resolution is worse by the factor $(300)^{1/2}/(26.4)^{1/2} = 3.37$ giving a resolution (Res) equal to

$$\text{Res} = \frac{128}{E^{1/2}} \qquad (4\text{-}9)$$

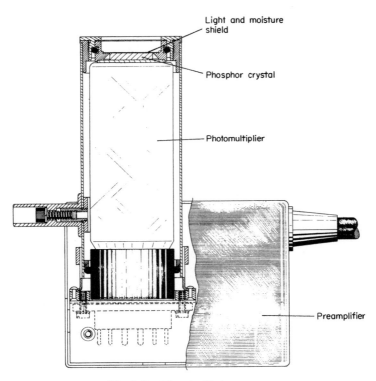

Fig. 4-13   The scintillation counter

Figure 4-13 shows a schematic diagram of the scintillation detector. The phosphor crystal is about two millimeters thick and is mounted in a light and moisture tight cover. This is necessary both to stop stray light from entering the photomultiplier and to prevent moisture reaching the rather hygroscopic sodium iodide crystal. These shields greatly reduce the sensitivity of the scintillation detector for the longer wavelengths and the optimum sensitivity range is from about 0.2 to 2 Å. This long wavelength limit overlaps somewhat with the useful short wavelength limit of the gas flow proportional detector (about 1.5 Å), therefore it is common practice to mount the two detectors in tandem, see Fig. 4-14. By allowing pulses to be collected from either or both detectors, optimum sensitivity can be achieved over most of the 0.2–20 Å range.

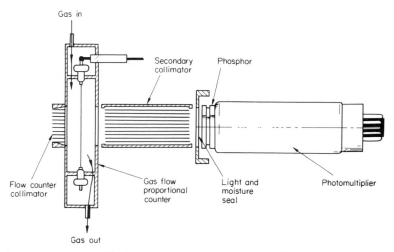

Fig. 4-14   The tandem detector assembly

The dead time of the scintillation detector (0.2 μs) is somewhat shorter than that of the gas flow proportional detector (0.5 μs), but this difference is to a large extent masked by the pulse counting circuitry where additional dead time—arising mainly from the pulse shaper—increases the effective dead time of the scintillation detector to about 1.2 μs and the gas flow proportional detector to about 1.6 μs.

Escape peak phenomena are also observed in the scintillation detector and in this instance it occurs when the incident X-ray photon ejects a K or L electron from the iodine atoms in the phosphor. However, the iodine L edges are around 5 keV and outside the normal working range of the detector. The iodine K edge is at 33 keV and for X-ray photons in excess of this energy level, escape peaks will occur.

4.2.3   The semiconductor detector

Semiconductor detectors operate by the creation, by incident radiation, of electron–hole pairs in the depletion layer of a reversed bias p-i-n junction.[28–31] The detector itself is a piece of p-type silicon which has been "drifted" with lithium atoms in order to increase the sensitive, or "intrinsic", layer. The p-type silicon is a semiconductor and the effect of the lithium is to compensate for charge which might be carried by impurities, which are inevitably present in the silicon single crystal. Figure 4-15 shows an idealized p-i-n junction consisting of a p-type layer, an intrinsic layer of any remaining free charge carriers, and the electric field strength is virtually constant over the whole of the depletion layer. Figure 4-16 shows a schematic representation of the Si (Li) detector and it will be seen that the sensitive layer is typically 3–4 mm thick. Unfortunately, the

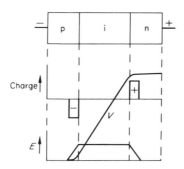

Fig. 4-15 A p-i-n junction showing idealized space charge, potential and electrical field strength distribution

lithium ions are still quite mobile at room temperature and it is generally necessary to keep the detector and its f.e.t. (field effect transistor) under cryogenic conditions. Liquid nitrogen is commonly used as a coolant and a moderately sized Dewar flask will hold sufficient to maintain the detector for several days. The f.e.t. is an integral part of the preamplifier and this is also cooled to minimize effects of electronic noise.

An X-ray photon entering the detector produces a "cloud" of electron–hole pairs and the number n of these pairs is proportional to the energy E of the incident photon, or

$$n = \frac{E}{\varepsilon} \qquad (4\text{-}10)$$

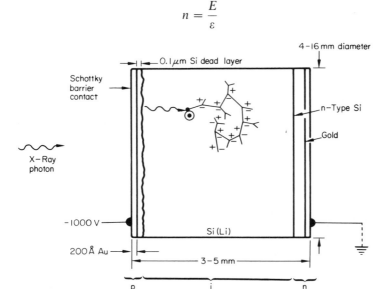

Fig. 4-16 A schematic representation of the Si (Li) detector diode and the X-ray photon interaction. (After Gedecke, D. A., *X-ray Spectrom.* **1**, 129 (1972))

where ε is the average energy required to produce one hole–electron pair. For a Si(Li) detector, ε has a value of 3.8 eV. In comparison with the gas flow proportional and scintillation counters, the number of primary electrons released per absorbed X-ray photon is considerably more and the resolution of the semiconductor detector is subsequently superior. The preamplifier noise contribution is a significant factor in the overall semiconductor detector resolution Γ, and the full width of an energy peak at one half its maximum intensity (FWHM) is given by

$$\Gamma = \{(\sigma_{\text{noise}})^2 + (2.35(\varepsilon FE)^{1/2})^2\}^{1/2} \tag{4-11}$$

σ_{noise} is typically 100 eV. F is the Fano factor which is about 0.125 at the current stage of detector technology. Figure 4-17 shows a curve of the overall detector resolution as a function of photon energy E. It will be seen that, for the noise level quoted, as $E \rightarrow 0$, $\Gamma \rightarrow 100$ eV. The amplifier noise is dependent mainly upon the amplifier time constant which in turn varies with the useful area of the detector surface. Typical detector areas are of the order of 12 mm^2 and the detectors generally have noise levels between 80–120 eV.[31]

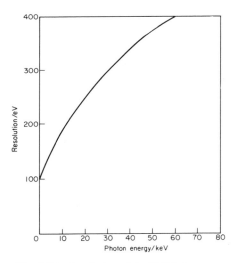

Fig. 4-17   Resolution of the Si (Li) detector

Figure 4-18 compares the relative resolutions of the gas flow proportional and scintillation counters with that of the Si(Li) detector. Also shown on the curve is the resolution of a typical crystal spectrometer. It will be seen that the resolution of the Si(Li) detector is better than that of the crystal spectrometer for wavelengths shorter than about 0.7 Å and the rapid development of energy dispersion spectrometers in the late 1960's came mainly as a result of this high inherent detector resolution.

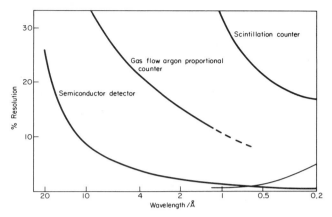

Fig. 4-18   Comparison of the resolutions of flow, scintillation and semiconductor counters, and the crystal spectrometer

For crystal dispersion spectrometers, the semiconductor detector offers little more than can be achieved with the gas flow proportional and scintillation counters. Since the semiconductor has to be operated under cryogenic conditions, it is less convenient to use than the other two detector systems and its use in X-ray spectrometry is, therefore, limited almost exclusively to that of a combined detection/dispersion device in the energy dispersion spectrometer. One of the more important requirements of this instrument is the need to be able to work at relatively high counting rates without significant loss of energy resolution. This is particularly critical where the detector is the only means of separating what might be a relatively small *wanted* signal, i.e., from the element being sought, from a large *total* signal, i.e., the total radiation flux from the specimen falling onto the detector. If the detector is unable to handle high counting rates, this will severely limit the allowable excitation of the specimen, with the consequence that detection limits will be poor. Total count rates up to around 70,000 counts/s can be conveniently handled at the present time, although it is to be expected that this will increase considerably over the next few years.

4.2.4   Pulse height selection

All of the detector systems employed in the modern X-ray spectrometer are proportional detectors. A proportional detector is one in which the energy of the incident X-ray photon determines the size of the voltage pulse produced by the detector. Where this is the case, it is possible to employ a means of selecting only a narrow range of voltage pulses, and this may be useful in discriminating against unwanted radiation.

Figure 4-19 illustrates the distribution of voltage pulses which would be produced when three separate X-ray energies were incident on a counter at a given time. In practice, this situation could easily occur when the first order reflection of wavelength λ_1, the second order of λ_2 and third order of λ_3 are all excited at the same time, and where $\lambda_1 = \lambda_2/2 = \lambda_3/3$. For example, first order phosphorus Kα ($\lambda = 6.16$ Å), second order calcium Kβ ($\lambda = 3.09$ Å) and third order gadolinium Lα ($\lambda = 2.05$ Å) since each of these will satisfy the Bragg Law giving nominally the same diffraction angle when measured with the same analysing crystal. The three different wavelengths will give rise to three different *average* voltage levels V_A, V_B and V_C corresponding to λ_1, λ_2 and λ_3 respectively. It has already been shown that the detector output pulses corresponding to a given photon energy will not all be of exactly the same size but will be distributed about a mean—the width of the distribution at half maximum being determined by the resolution of the detector. An acceptance window can be used to reject all voltage levels falling outside of the limits to which it is set and in Fig. 4-19 only the pulses of average size V_B will be collected. The acceptance window is usually defined at the lower voltage level by an adjustable "threshold" value, and at the upper level by an adjustable window which is simply the difference between upper and lower levels.

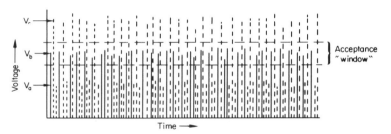

Fig. 4-19   Removal of unwanted pulses by means of pulse height selection

Pulse height selection may also be useful for reducing background particularly in the longer wavelength regions. The majority of the background radiation arises from scatter of the primary radiation from the X-ray tube. This radiation is diffracted by the analysing crystal and so it is to be expected that the average order of the background increases with the wavelength of the measured radiation. This is illustrated in Fig. 4-20, which shows the analysis of background arising from a sample of distilled water when a chromium anode X-ray tube was used as the excitation source. The analysis was made with a 400-channel analyser attached to the linear amplifier of a detector. It will be seen that in the short wavelength region the wavelength of the background is exclusively first order, hence an analysis line in this region will always be superimposed on background of identical wavelength. This renders the pulse height selection completely ineffective, since due to the relatively poor resolution of the detector, a wavelength difference of at least 20 % would be required for complete separation.

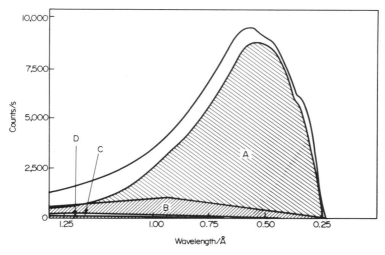

Fig. 4-20   Variation in the order of scattered background
A : First order
B : Second order
C : Third order
D : Fourth order

At longer wavelengths, however, the background becomes of increasingly higher order, for example, at 1.25 Å less than 50 % is first order, hence, approximately half of the background can be removed by careful application of pulse height selection. For the very long wavelengths such as aluminium Kα (8.34 Å) and magnesium Kα (9.89 Å), the background may be as much as sixth or seventh order and almost complete removal of scattered background is possible.

The system just described is called a "single" channel pulse height selector and its major use is as an aid to the crystal spectrometer in removing unwanted radiation. In energy dispersion spectrometers, a complete voltage pulse distribution is required and for this purpose a "multi"-channel pulse height analyser is used. A multichannel pulse height analyser can be considered to be a large number (usually several thousand) of single channel pulse height selectors. In practice, this may be a hard-wired instrument designed specifically for the purpose, or a section of the core of a digital computer where individual voltage levels are stored in different memory locations, after analogue (pulse height) to digital conversion.

4.3   SEPARATION OF A POLYCHROMATIC BEAM OF RADIATION

A spectrometer is a device which splits a polychromatic beam of radiation into its component wavelengths; its ability to do this is usually expressed in terms of its "dispersion". A glass prism is a very simple optical spectrometer and most readers will be familiar with the mechanism by which white light is broken up, or dispersed, into the seven principal colours. A further common example of a spectral separation is the formation of a rainbow produced by the combined effects of refraction and internal reflection of sunlight by drops of rain. Unfortunately, we are unable to utilize these phenomena to disperse X-radiation, since the index of refraction for X-rays is very close to unity (0.99999) and total external reflection occurs only below about 0.25°. It has, however, already been shown in Section 3.3 (p. 46), that the property of diffraction can be used to separate a polychromatic beam, in this instance, the diffracting medium being a single crystal or a grating.

The crystal spectrometer has for many years been the standard means of dispersion in analytical X-ray instrumentation, but is by no means the only way of separating a given wavelength from a polychromatic beam. In many instances, a good alternative is the so-called "energy dispersive" technique, and the application of this method has increased steadily since the late 1960's. The energy dispersive method is based on the use of a proportional detector and a multichannel analyser. A critical feature of a proportional detector is that the energy of an X-ray photon entering the detector determines the size of the voltage pulse produced. Thus a polychromatic beam of radiation incident upon the detector will produce a "spectrum" of voltage pulses, which have a size distribution proportional to that of the energy distribution of the incident polychromatic beam. A multichannel analyser is used to separate the spectrum of voltage levels into narrow voltage bands, thus allowing the measurement of individual energies.

Since the analytical X-ray region lies between $0.2 - 20$ Å, both crystal and energy dispersion spectrometer systems are designed to give optimum performance over this range. As will be seen, the crystal spectrometer is generally best towards the long wave end of this analytical range and the energy dispersion spectrometer is more useful for the measurement of shorter wavelengths.

4.3.1 The crystal spectrometer

In the crystal spectrometer the measured parameter is the diffraction angle and a useful means of quantifying the ability of a given crystal to separate two wavelengths λ_1 and λ_2 is the "angular dispersion", $d\theta/d\lambda$, of the crystal. This represents the angular difference $d\theta$ obtained from the wavelength difference $d\lambda$. $d\lambda$ is simply the numerical difference between λ_1 and λ_2 in Ångstrom units and $d\theta$ the difference, *in radian measure*, between the diffraction angles of λ_1 and λ_2. $d\theta/d\lambda$ can be derived by differentiating λ with respect to θ in the Bragg relationship and taking the reciprocal of the result

$$\lambda = \frac{2d}{n} \sin \theta \qquad (4\text{-}12)$$

or

$$\frac{d\theta}{d\lambda} = \frac{n}{2d} \frac{1}{\cos \theta} \qquad (4\text{-}13)$$

Angular dispersion should not be confused with resolving power or resolution. In light optics two points are said to be "resolved" if their corresponding diffraction patterns are sufficiently separated to be distinguished. The numerical ability of a system to resolve two points is its "resolving power", $\lambda/d\lambda$. The resolving power can be derived by combining equations (4-12) and (4-13)

$$\frac{\lambda}{d\lambda} = \frac{1}{d\theta \cot \theta} \tag{4-14}$$

It should be noted that angular dispersion increases with increase in the order of reflection n, or with decrease in the $2d$ value of the analysing crystal. Resolving power also increases with the diffraction angle but since for a given wavelength the diffraction angle is directly dependent upon n and $2d$, angular dispersion is a more useful expression in practice.

The expression "resolution" is often used to convey the ability of a total system to separate lines. Figure 4-21 shows three schematic line pairs each separated by the same absolute angular dispersion. Only the profiles shown in (a) are completely separated or resolved.

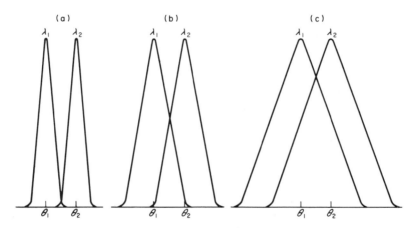

Fig. 4-21   Effect of line breadth on peak separation

Hence angular dispersion *alone* is not sufficient to define whether or not peaks are separated. It is apparent that profile shape, generally defined by peak width at half height, is also a critical parameter. It will also be seen that if an observation were made at θ_1 in (b) only λ_1 would be measured, since the λ_2 component at θ_1 is zero. For this reason the peak width at half height is often used as a measure of the resolution of a system. However, this is only true in the idealized case where the profiles are symmetrical and base widths are exactly twice the widths at half height.

4.3.2   Geometry of the crystal spectrometer

Many types of geometric arrangements have been employed for the X-ray spectrometer[32–35] including use of crystals in flat, curved, logarithmic and transmission configurations. Spectrometers fall into two broad categories, namely sequential (scanning) and simultaneous (multichannel). In the sequential type, a single spectrometer is scanned over a selected angular range, using a continuous movement for qualitative work and incremental movement between specified angles for quantitative work. In the multichannel system, many (generally 6 to 30) fixed spectrometers are grouped around the specimen giving simultaneous readings on every element for which a spectrometer has been provided. Flat crystal optics are generally better suited for scanning spectrometers where a single crystal must give optimum performance over a relatively wide wavelength range. In multichannel spectrometers, the optimum geometry can be selected for each wavelength and in this case curved or logarithmic crystal optics are generally the best.

Figure 4-22 shows a schematic diagram of a flat crystal spectrometer and the optical path includes an analysing crystal, a primary collimator between specimen and crystal, a secondary (auxiliary) collimator between flow and

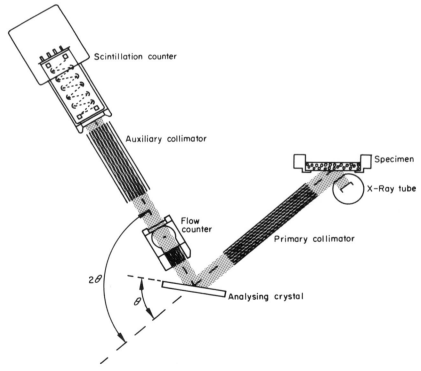

Fig. 4-22   The flat crystal spectrometer

scintillation counter and the means of rotating the detector combination at twice the angular speed of the analysing crystal. In a scanning system this latter movement is controlled by the goniometer which is a precise mechanical device capable of reproducing the $2:1$ coupling and absolute value of 2θ to better than $\pm 0.01°2\theta$. Figure 4-23 shows an exploded view of an actual spectrometer and this will serve to indicate the additional details of a typical system.

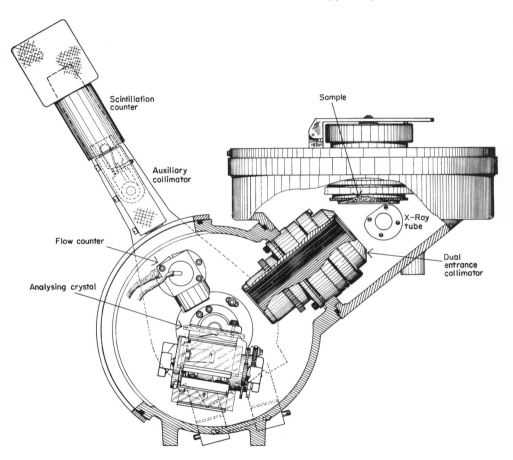

Fig. 4-23   Schematic diagram of the Philips PW1410 X-ray spectrometer

The first thing to note is that both analysing crystal and collimator assemblies allow a choice to be made. It is usual to provide two primary collimators and between two and six analysing crystals. The collimators consist of sets of parallel blades, usually with spacings of about 150 μm (fine) and 450 μm (coarse). The two sets are concentrically mounted in a cylinder and one or the other set can be brought into the optical path by rotating the cylinder. The crystals are mounted either on a translatable sledge or on a rotatable drum, and selection of a given crystal is achieved by a mechanical movement coupled to the crystal mount. Two detectors are provided: a gas flow proportional counter which is most sensitive in the 20–2 Å region, and a scintillation counter which is best suited for the measurement of wavelengths in the 2–0.2 Å region. Figure 4-24 indicates that the absorption of X-radiation by air also becomes important at around 2 Å and for this reason the gas flow proportional counter is mounted inside the main chamber of the spectrometer, which is then evacuated. Multiple sample presentation facilities are generally provided and some form of air lock is incorporated for loading specimens. Where specimens in the liquid phase are to be analysed, a helium flushing system is used. Figure 4-24 also shows the transmission curve for helium and it will be seen that this is satisfactory for all but the very long wavelengths.

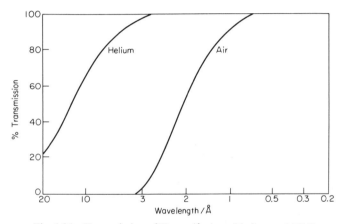

Fig. 4-24   Transmission of X-rays in air and helium at N.T.P.

Figure 4-25 shows the way in which the average resolution of the spectrometer varies over the analytical wavelength range. Also shown on the curve is the percentage difference between neighboring atomic numbers for Kα wavelengths. In practice there are of course more lines to resolve than just the Kα's, but Fig. 4-25 gives at least an indication of the resolution requirements of the spectrometer. Study of the two curves shows that the major problems of resolution are likely (but not always) to occur in the shorter wavelength region. For this reason, extra resolution is provided for this region in the form of an auxiliary collimator mounted in front of the scintillation counter. Since extra collimation can only provide better resolution at the expense of loss of intensity, mounting the auxiliary collimator in front of the scintillation counter, rather than in front of the gas flow proportional counter, ensures high resolution at short wavelengths without additional intensity loss at longer wavelengths where intensity is generally at a premium.

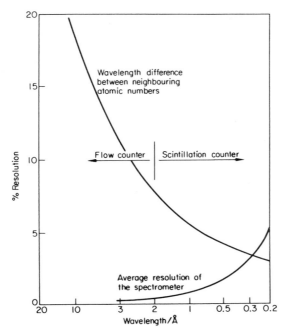

Fig. 4-25   Variation of spectrometer resolution with wavelength

4.3.3   Resolution of the spectrometer

As has already been intimated, the ultimate resolution of the spectrometer will depend upon the angular dispersion of the crystal and the shapes of the line profiles. The angular dispersion has already been shown to be directly related to the "d" spacing of the analysing crystal and "n" the order of the reflection. Equation (4-13) shows that angular dispersion increases with increase of "n" and decreases with increase in "$2d$". As an example, Fig. 4-26 shows the barium K spectrum recorded with LiF (200) crystal, $2d = 4.028$ Å. The spectrum has been measured over a sufficiently wide range to include first, second and third order reflections, i.e., $n = 1$, $n = 2$, and $n = 3$. As the value of n increases so does the separation of the lines. Thus whereas for the first order, the α_1 and α_2 lines are completely overlapping, they are partially resolved in the second order and completely resolved in the third. A similar circumstance is apparent for the $\beta_{1,3}$, β_2 doublet. Figure 4-27 shows the effect of the $2d$ spacing and illustrates the molybdenum $K\alpha_{1,2}$ doublet recorded with different cuts of lithium fluoride.[35] Here the $2d$ value varies from 4.028 Å for the (200) sets of planes to 1.645 Å for the (422) sets of planes. The increased peak separation with decrease in $2d$ is clearly seen. Note that increasing the angular dispersion, either by increase in n or decrease in $2d$, leads in both cases to a lower absolute intensity.

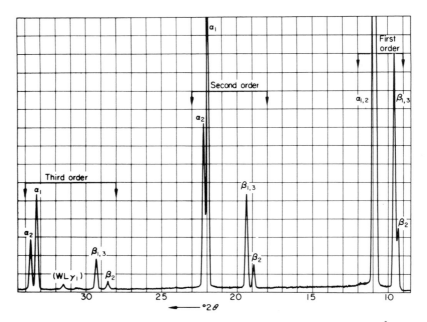

Fig. 4-26   Barium K spectrum recorded with a LiF(200) crystal 2d = 4.028 Å

The shape of the line profile is dependent upon several factors of which the most important are the divergences allowed by the primary and secondary collimators, and the so-called "mosaic" spread of the analysing crystal. Figure 4-28 illustrates the divergence allowed by the primary collimator. Here a collimator is shown of length l and spacing s, and it is clear that not only will the collimator allow the direct passage of rays to the crystal, making an angle θ_i, but it will also pass rays with a maximum divergence of α, equal to arctan (s/l) (radians). However, if the crystal is perfect, i.e., all the reflecting planes lie exactly parallel to the surface, only the direct rays through the collimator will be diffracted, since only for these direct rays does the angle of incidence θ_i equal the angle of reflection θ_r. For diverging rays the angle of incidence varies over the range $\theta_i \pm \alpha$. These diverging rays can, however, be made to satisfy the Bragg condition by changing the angle of the crystal by an amount $\pm \delta\theta$ where $\delta\theta = \alpha/2$. This gives angles of incidence and reflection of

$$\theta_i - \delta\theta = \theta_r + \delta\theta$$

and

$$\theta_i + \delta\theta = \theta_r - \delta\theta$$

for the two extremes of divergence. Assuming a homogeneous distribution of intensity from the specimen one pair of collimator blades would give rise to the diffracted intensity distribution shown in Fig. 4-28, viz. a triangular shape of

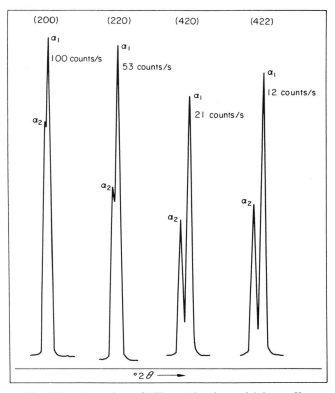

Fig. 4-27 Comparison of LiF crystals using molybdenum Kα

Reflection planes	2d/Å	Relative intensity (Mo Kα)	dθ in ° 2θ for M₀ Kα₁,₂
(200)	4.028	100	0.60
(220)	2.848	53	0.66
(420)	1.802	21	1.39
(422)	1.645	11.2	1.53

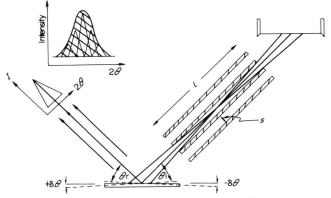

Fig. 4-28 Effect of primary collimator on profile shape

base 2α. Since many pairs of collimator blades are present (usually at least 10) the total shape of the profile is made up of many triangles of decreasing height from the centre of the profile. Decreasing height because the further from the pivot point of the crystal, the greater the defocussing due to $\delta\theta$.

In practice it is most unlikely and indeed most undesirable, that the crystal be perfect. Most undesirable, because a perfect crystal will give rise to extinction effects similar to that shown in Fig. 4-29. Sketch (a) illustrates the case of a perfect crystal shown to be made up of blocks each perfectly orientated with respect to one another. Incident rays falling onto the crystal surface at angle θ are diffracted at angle θ in the normal way. However, diffracted radiation from a given lower block also makes an angle θ with the underside of the corresponding upper block, so diffraction can occur from the underside of this upper block with subsequent loss of total diffracted intensity in the direction of the detector.

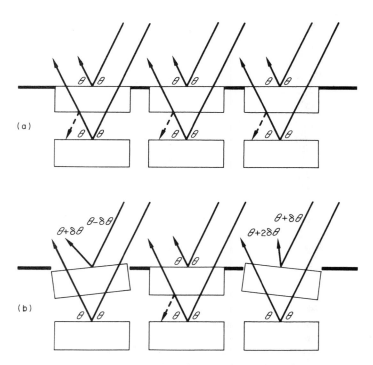

Fig. 4-29   Extinction effects in perfect and "ideally imperfect" crystals

Sketch (b) illustrates the case of an "ideally imperfect" crystal where the surface blocks are mis-orientated with respect to the lower blocks. In this instance, diffraction loss is far less because at a given 2θ value, the Bragg condition is seldom met by rays falling onto the undersides of the upper blocks. It will also be seen that the crystal as a whole will diffract over a range of 2θ values (i.e., the profile will be broadened) since there is a mis-orientation of the upper blocks relative to the average surface of the crystal. Thus by deliberately making the surface of the crystal more "mosaic" by abrasion, the total diffracted intensity can be increased but only at the expense of a broadened diffracted profile. Crystals are generally abraided by a variety of surface treatments including ultrasonic treatment, flexing, and rubbing with abrasive polishes.[36,37]

Figure 4-30 illustrates the effect of the mosaic nature of the crystal on the geometry of the total system. As before, rays which are parallel to the blades of the primary collimator are diffracted by crystallites parallel to the average surface of the crystal. However, in this instance, the divergent rays from the primary collimator can also be diffracted, in this case by crystallites which are misorientated with respect to the average surface of the crystal. The result is that far more incident rays are "collected" by the crystal, with a subsequent increase in diffracted intensity. However, since most of the additional diffracted intensity arises from divergent rays, the crystal will diffract over a relatively wide range of θ with a significant broadening of the diffraction profile (lower profile in Fig. 4-30). The effect of interposing the secondary collimator between crystal and detector is to prevent the very divergent diffracted rays from entering the detector. The resulting line profile (upper profile in Fig. 4-30) is of lower intensity but has a much narrower base than the profile coming from the crystal.

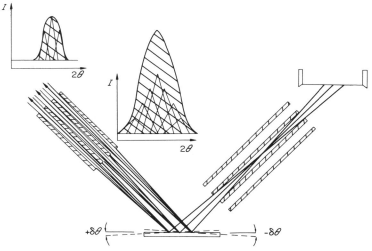

Fig. 4-30 Effects of primary and secondary collimation on profile shape

In general terms it can be stated that the width of a profile at half height W is determined by convoluting the divergence analysing crystal W_x or

$$W = (B_c^2 + W_x^2)^{1/2} \qquad (4\text{-}15)$$

It has already been stated that B_c is equal to arctan (s/l) and typical values for a 10 cm long primary collimator are given in Table 4-4. W_x generally has a value of

TABLE 4-4.
Divergence allowed by a 10 cm
parallel plate primary collimator

Spacing/μm	Angular divergence/degrees
150	0.086
450	0.258
900	0.516

around 0.2 although it can vary from as little as 0.1 for natural cleaved crystals, to 0.6 for very poorly orientated crystals such as pyrolytic graphite.

Figure 4-31 shows the second order Kα spectrum of copper recorded using a quartz crystal ($W_x \simeq 0.1$) and a pyrolytic graphite crystal ($W_x \simeq 0.6$) utilizing

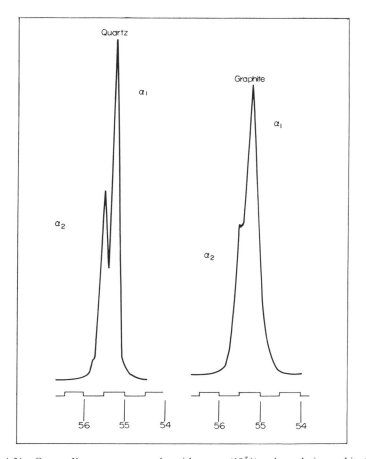

Fig. 4-31   Copper Kα spectrum recorder with quartz (10$\bar{1}$1) and pyrolytic graphite (002)

a conventional powder diffractometer. The superior resolution of the quartz crystal is clearly shown, although in this experiment the effective divergence due to the slits of the diffractometer is the limiting factor. It should also be noted that where line multiplets are involved, and this is generally the case, the effectiveness of the spectrometer in separating similar wavelengths from different elements may well be limited by the absolute angular dispersion of the multiplets. For example, in Fig. 4-31 although the peak width at half height of the $K\alpha_1$ line recorded with quartz is only about $0.2°2\theta$, the actual width of the α_1/α_2 doublet is $0.6°2\theta$—in fact, exactly the same as the profile obtained with pyrolytic graphite.

It will be apparent from equation (4-15) that there is an incentive in matching the values of B_c and W_x in order to ensure that one or the other is not the limiting factor in determining W. For example, it would be foolish to employ a 900 μm collimator ($B_c = 0.516$) along with a topaz crystal ($W_x = 0.08$) since the too wide collimator would allow additional background radiation to enter the detector without an increase in the signal from the measured wavelength. Similarly, the use of a 150 μm collimator ($B_c = 0.086$) with a pyrolytic graphite crystal ($W_x = 0.6$) would cause a great loss in intensity without improvement in resolution.

4.3.4 Choice of analysing crystal

The usefulness of an analysing crystal is governed to a large extent by two basic properties, its angular dispersion (p. 76) and its reflecting power (p. 85). Although other features such as temperature stability, crystal fluorescence, missing reflections and/or forbidden reflections may all be important,[32] they are rarely critical and a crystal invariably stands or falls on its angular dispersion and/or its reflecting power. This is even more true today than in the early days of X-ray spectrometry, since the trend has always been to produce faster analyses without loss of accuracy, which in turn requires higher intensities. Generally speaking, higher intensities can always be obtained at the expense of peak separation, hence the importance of the basic properties of the analysing crystal.

Since the range of the spectrometer is 0.2–20 Å, it will be apparent from the Bragg Law, equation (4-12), that crystals of at least 20 Å $2d$ spacing must be available in order to diffract the longest wavelengths. In practice, the maximum obtainable value of θ is 75°, thus the maximum value that $\sin \theta$ can take is around 0.95. Thus a crystal of $2d$ equal to about 21 Å would cover the whole of the wavelength range. Unfortunately, however, the angular dispersion of such a crystal for the shorter wavelengths would be completely unacceptable because of its large $2d$ spacing. A compromise obviously has to be made since whereas a large $2d$ spacing is required to cover a large wavelength range, as small a $2d$ value as possible may be required to give the necessary angular dispersion. For this reason, a range of analysing crystals is nearly always employed to adequately cover the whole wavelength range. There is a continuing search for new analysing crystal.[35] and extensive lists of crystals and their properties are available.[39] Table 4-5 lists the more commonly used crystals and represents a "state of the art" list as of going to press.

TABLE 4-5.
Commonly used analysing crystals

Crystal	Reflection plane	$2d/\text{Å}$	Usual atomic number range
Lithium fluoride (LiF)	(420)	1.802	high resolution, short wavelength work
Lithium fluoride (LiF)	(200)	4.028	all $Z > 19$
Germanium (Ge)	(111)	6.532	gives very low second order intensity, useful for harmonic overlap problems
Pyrolytic graphite (PG)	(002)	6.715	P(15), S(16) and Cl(17)
Pentaerythritol (PE)	(002)	8.742	Al(13) through K(19)
Ammonium dihydrogen phosphate (ADP)†	(101)	10.64	Mg(12)
Rubidium acid phthalate (RAP)	(001)	26.1	F(9) and Na(11)

† At the time of going to press there is significant evidence to suggest that ADP will soon be replaced by sorbitol hexa-acetate (SHA).[7]

4.3.5 Angular reproducibility of the spectrometer

It is generally required to reproduce the diffraction angle to better than $\pm 0.02°2\theta$, hence the goniometer is necessarily a fairly expensive piece of precise mechanical engineering. Although for most purposes temperature stabilization is unnecessary from the point of view of the gears and moving parts of the goniometer itself, it may well be required to reduce the thermal expansion effects of the crystal, since a small change in the linear "d" spacing is greatly amplified at large values of 2θ. If $2d/n$ in equation (4-12) is replaced by D, then

$$D = \frac{\lambda}{\sin\theta}$$

and

$$\frac{d(D)}{d\theta} = \frac{-D^2\cos\theta}{\lambda} \tag{4-16}$$

Equation (4-16) expresses the angular change $d\theta$ resulting from a change in $2d$ spacing $d(D)$. $d(D)$ per deg C is simply the thermal expansion coefficient *for the reflection planes in question*. Of the common crystals, pentaerythritol has by far the largest thermal expansion coefficient being equal in this instance (for the (002) reflection) to 0.0013 Å per deg C. Substitution of data for Al Kα radiation ($\lambda = 8.34$ Å) in equation (4-16) gives

$$d\theta = -\frac{0.0013 \times 8.34 \times 57.3}{(8.742)^2 \times 0.309} = -0.026° \text{ per deg C}$$

or in terms of d(2θ), an angular change of $-0.052°$ per deg C. A similar calculation for Si Kα radiation ($\lambda = 7.13$ Å) gives an angular change of $-0.023°2\theta$ per deg C and for most analytical work these are the only two element crystal combinations which give significant temperature effects.[40]

4.3.6 Use of primary beam filters

In the early days of X-ray spectrometry when the ratings of X-ray tubes and high voltage generators were relatively low, there was a considerable incentive to select an X-ray tube which gave optimum excitation for a certain wavelength range. With the advent of higher powered X-ray tubes and generators, the need for tube changing has decreased, the general exception being where the element from which the X-ray tube anode is made is to be determined in the sample. An example of this would be the determination of chromium or manganese with a chromium anode X-ray tube. A possible means of avoiding this problem is by the use of an aluminium filter placed between X-ray tube and sample which filters out the portion of the tube spectrum containing the characteristic tube lines. Figure 4-32 shows the effect of the primary beam filter on

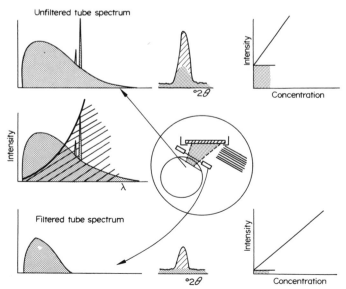

Fig. 4-32   Effect of the primary beam filter

both the primary spectrum from the X-ray tube and the excited characteristic radiation from the sample. It will be noted, however, that not only does the filter remove the characteristic lines from the tube spectrum (thus obviating the high "blank" reading at the measured wavelength) but it also removes the portion of the primary spectrum that is most efficient in exciting the required wavelength in the sample. It is found, therefore, that not only is the true background reduced when the filter is employed, but the counts/s per % for the excited element also drops.

From the foregoing, it will be apparent that thickness of the filter is quite critical, since increasing the thickness reduces both background and peak above background. In this instance, in order to establish the optimum thickness, series of measurements were done on a chromium containing steel utilizing different thicknesses of aluminium on the primary beam filter. Figures of merit were then calculated and the result is shown in Fig. 4-33. Below 200 μm, the reduction in the background is the most critical and from 50 to 200 μm the figure of merit curve increases steadily. Above this thickness, however, the attenuation of the peak counting rate becomes more important and the figure of merit begins to fall off.

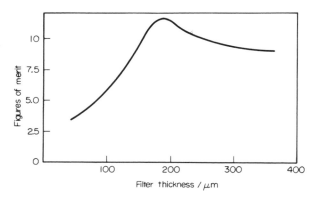

Fig. 4-33   Effect of primary beam filter thickness on figure of merit for Cr Kα excitation: Cr anode tube at 50 kV 40 mA; 150 μm primary collimator; Flow counter; LiF (200) crystal

The situation of having to choose between different sets of instrument variables is fairly common in X-ray spectrometry and a "figure of merit" is generally employed in order to establish the optimum condition. Section 6.2 (p. 110) discusses the expression relating the percentage counting error ε %, with the peak I_p and background I_b counting rates. It will be seen from equation (6-5) that for a fixed total analysis time T, ε % will be minimum when $I_p^{1/2} - I_b^{1/2}$ is at a maximum. For this reason, $I_p^{1/2} - I_b^{1/2}$ is generally taken as the figure of merit.

4.3.7 Typical instrumentation for analysis with a crystal spectrometer

A typical instrumental set-up for work with a crystal spectrometer includes three major portions, the source, including stabilized high voltage generator and X-ray tube; the spectrometer itself, including specimen handling facilities, and the detection and output device. Figure 4-34 shows these various components in block diagram form. The spectrometer is generally designed to operate under analogue or digital output conditions. The analogue mode is used for qualitative analysis and in this mode pulses from the detector are integrated in a ratemeter circuit, the function of which is to integrate the digital pulses for a selectable time which is determined by the "time constant" of the rate meter.

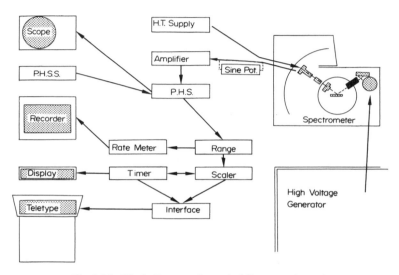

Fig. 4-34 Block diagram of a typical X-ray spectrometer

The resulting analogue output is then displayed on the recorder. By coupling the scanning speed of the goniometer to the chart speed of the recorder, a recording can be made of counting rate as a function of diffraction angle. For example, the chart shown in Fig. 4-26 was obtained by this means. A range selector is also made available to ensure that the required signal falls within the range of the recorder.

The digital output mode is generally employed for quantitative analysis and here pulses from the detector are counted by a scaler which is in turn coupled to a timer. Two modes of counting are possible, fixed time and fixed count. In the fixed time method of counting, a time is selected on the timer and the counting sequence started. This starting sequence starts the scaler counting and the timer timing at the same instant, and both continue until the selected time has been reached. At this point, the timer sends a stop pulse to the scaler which then displays the counts N obtained in the selected time T. The counting rate I is simply N/T. In the fixed count method, a certain number of counts is preselected on the scaler. In this instance, the timer is stopped by a pulse from the scaler when the preset number of counts is reached. The timer then displays the time required to obtain the selected number of counts. As before, the counting rate is obtained by dividing counts by time.

The output device may simply be a visual display of the contents of the scaler and timer, or there may be some means of recording the time and count data on a printer or teleprinter. It may also be useful to interface the timer and scaler straight to a small computer which can in turn manipulate the data to give direct concentration output.

In addition to the rate meter and scaler/timer circuitry, the electronic circuit panel will probably contain modules for pulse height selection (P.H.S.), with perhaps a means of scanning through the pulse height range (P.H.S.S.) plus some form of pulse monitor (usually a scope). Modern scanning spectrometers usually provide the means of automatically applying pulse height selection and this is generally done by attenuating the pulses with a sinusoidal potentiometer coupled to the goniometer.

4.3.8   The energy dispersion spectrometer [31,41–43]

In the energy dispersion spectrometer, the measured parameter is the size of the voltage pulse from the detector, this being proportional to the energy of the measured X-ray photon. The critical factor in assessing the ability of an energy dispersion spectrometer to separate one energy from another is its resolution Γ and this has been shown to be related to energy E by an equation of the form (see equation (4-11))

$$\Gamma = (K_1 + K_2 E)^{1/2} \tag{4-17}$$

where K_1 is dependent only on the noise component of the total system and K_2 is related to the true statistical fluctuation of the Si(Li) diode. Unlike the crystal spectrometer, the resolution of the energy dispersion spectrometer *improves* with decrease of wavelength as is shown in Fig. 4-18. This makes the energy dispersion spectrometer particularly well suited for work with shorter wavelengths. Figure 4-35 shows a diagrammatic representation of a typical energy dispersion system. The sample is irradiated by miniature X-ray tube or radioisotope source and the excited fluorescence radiation collected by the Si(Li) detector. The detector consists of a Si(Li) diode, and this, along with a charge-sensitive preamplifier, is cooled with liquid nitrogen. Pulses from the detector are amplified and analysis is performed by the multichannel analyser, which may be a hardwired system or a suitably programmed digital computer. Whichever of these two configurations is employed there is generally provision for internal calibration and spectrum stripping, in addition to the spectrum analysis capability. The output device is generally a display tube and/or printer.

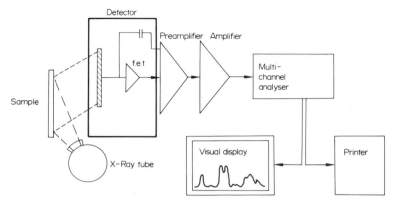

Fig. 4-35   The energy dispersion spectrometer

4.3.9   Sensitivity of the energy dispersion spectrometer

The sensitivity of the energy dispersion spectrometer is determined largely by the efficiency of the excitation source and the quantum counting efficiency of the detector. Sources may be radioisotopes of around 10–30 MCi activity, miniature X-ray tubes operating at about five watts, or secondary fluorescer sources. The radioisotope generally offers the advantages of high short term (or even long term depending on the half-life) stability, low cost, and small physical size. Against this, however, the total useful quantum yield is low and the range of excitation energies covered with a given source is smaller. Table 4-2 lists a few typical radioisotope sources which have been successfully employed. The miniature X-ray tube is more expensive and has a much greater physical size, but against this it gives generally more than sufficient quantum yield and can be operated up to 60 kV. This in turn gives most of the advantages offered by the conventional X-ray tubes described in Section 4-1 (pp. 52–58).

The tube shown in Fig. 4-36 is typical of those currently in use and is very similar to the type used in dental surgery. This particular tube is air-cooled and was designed for use in a single (selectable) channel analyser.[44] Anode materials used in this configuration include copper, silver and zirconium, and the tube has a maximum rating of 30 W (30 kV, 1 mA).

One of the more recent advances in the provision of sources for the energy dispersion spectrometer is the secondary fluorescer.[45] The secondary fluorescer is based on the use of a high intensity primary source to excite a selected element (the secondary fluorescer) which in turn excites the required wavelength, or wavelength range, from the specimen. The concept itself is not new and it is the basis of the $\gamma - X$ source already discussed in Section 4.1.5 (p. 60). Probably the greatest advantage to be offered by the secondary fluorescer source is that careful choice of the secondary fluorescer element will yield an almost monochromatic source, in turn giving very specific specimen element excitation

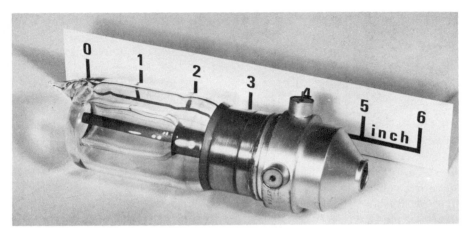

Fig. 4-36   The miniature X-ray tube. (After Gammage, C. F., *X-ray Spectrom.* **1**, 99 (1972))

with associated high sensitivity and low background. Added benefits include the reduction in the number of scattered primary source wavelengths, which are generally inevitable in the use of radioisotope sources, plus the freedom from difficulties involved with the handling of relatively high activity radioisotopes.

As discussed in Section 4.1.4 (p. 59), there is growing interest in proton excited X-ray spectrometry, particularly in view of its high potential sensitivity. Figure 4-37 shows the spectrum obtained from a 300 Å layer of Al_2O_3 using a 100 keV proton source. A special windowless detector was used for this application with a reported efficiency of about 3 % for oxygen Kα. Similar low atomic number

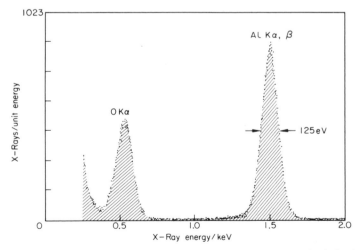

Fig. 4-37   Spectrum of Al_2O_3 obtained with a 100 keV proton source and windowless Si (Li) detector. (After Goulding, F. S. and Stone, Y., *Science* **170**, 280 (1970))

element data have been obtained using a special aluminium target X-ray tube.[46]

The quantum counting efficiency of the detector is an important factor in the overall sensitivity of the energy dispersion spectrometer, and Fig. 4-38 shows the variation of this parameter with wavelength for both Si(Li) and Ge(Li) detectors. The sensitivity fall-off at the long wavelength end of the range is due to several factors of which the more important are absorption by the beryllium window and the "dead" layer at the detector surface. The beryllium window is typically 7–50 μm thick and the curves shown in Fig. 4-38 are for a 12.5 μm window. The transmission for 20 Å radiation by this thickness is around 30% but even by complete removal of the window, the detector efficiency for 20 Å radiation cannot be improved to better than about 5% even for the Si(Li) detector. It is thus apparent that for the longer wavelengths, other factors come into play and the dead layer of the silicon diode is particularly critical. Figure 4-16 shows that a dead layer of about 0.1 μm exists at the detector surface and radiation must pass through this dead layer, plus the thin gold layer, before reaching the active volume of the counter. For 20 Å radiation a 0.1 μm layer of silicon ($\mu \simeq 7000$) would attenuate the signal by a factor of 2 with a similar attenuation by the 200 Å (i.e., 0.02 μm) layer of gold ($\mu \simeq 30,000$).

The sensitivity fall off at the short wavelength end of the range is due to the finite thickness of the diode itself. This thickness is typically 3–5 mm and the curves shown in Fig. 4-38 are for a 3 mm thick diode. The theoretical absorption

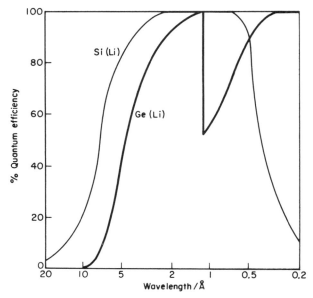

Fig. 4-38   Quantum counting efficiency curves for 3 mm thick Si(Li) and Ge(Li) detectors, 12.5 μm beryllium windows.

of a 3 mm thick piece of silicon for 0.2 Å radiation ($\mu \simeq 0.2$) is about 15% for radiation at normal incidence and this is of the same order as the experimentally determined value shown in Fig. 4-38. The critical depth for 0.2 Å radiation would be around 8 cm, which is currently impracticable for the lithium drift process.

Germanium, drifted with lithium, has also been used as a detector for energy dispersive spectrometry and with a critical thickness that is an order of magnitude less than that of silicon (μ Ge for 0.2 Å radiation is about 2) it is obviously better suited to the measurement of shorter wavelengths at least from the point of view of absorption. Unfortunately, the higher atomic number of germanium also increases the absorption of longer wavelengths in the dead layer, particularly for wavelengths close to the short wavelength side of the germanium K edge. Figure 4-38 also shows that the crossover point of the Si(Li) and Ge(Li) curves is about 0.5 Å, so the Si(Li) detector is generally better suited for the wavelength range in excess of this value. A further complication which arises when the Ge(Li) detector is employed over the 0.2 to 1 Å region is the occurrence of escape peak phenomena. These are produced when the incident X-radiation can excite germanium K radiation (λ_{edge} equal to 1.12 Å) and are similar to those produced in the gas flow proportional counter.

4.3.10   Data handling in the energy dispersion spectrometer

The most important part of the energy dispersion spectrometer—other than the detector itself—is the spectrum analyser. As previously mentioned, this is generally a hardwired multichannel analyser, specially designed for this purpose, or a digital computer. The great advantage to be offered by the use of the computer as a spectrum analyser is that it can also be used for further data handling, including conversion of "true" counting rates to concentration. Roughly, two thousand 16-bit words are sufficient for spectrum analysis, with a further thousand words for spectrum stripping. Intensity corrections of the type described in Section 7-9 (p. 132) can be handled within a thousand words of core for a reasonable large sample matrix (about 8×8 equations corresponding to 8 analysed elements and up to 8 corrections per element), thus the majority of what is required can be handled by a 4K, 16-bit word, digital computer. At the present stage of technology, the 4K digital computer is probably both cheaper and more versatile than a multichannel analyser with spectrum stripping capability.

Whichever system is employed, capabilities in addition to spectrum analysis and spectrum stripping are required before the count rates can be converted to concentrations. Among the corrections which may have to be applied are those for pulse pile up, energy resolution and baseline stability, and dead time losses.[31] The pulse shaping circuitry is not unlike that used in the gas flow proportional counter (see p. 65) and a compromise always has to be made between shortening the time constant of the pulse shaper to allow the handling of high count rates, and lengthening the time constant to maintain good detector resolution. Although baseline restoration is generally employed, this breaks down at high count rates where pulses tend to overlap giving an artificially high baseline. This effect is demonstrated in Fig. 4-34 where series of pulses are shown on a time scale. The width T of the pulse is determined by the time constant of the pulse shaper, and the pulse height E_A by the energy of incident radiation. In the lower part of the figure the pulses are resolved, but in the upper diagram they overlap each other since a second pulse starts to form before the first has decayed away. This sets up an artificial baseline E_B with a corresponding change in average pulse height of E_A to $E_A - E_B$. Obviously this effect will get worse as the pulse frequency, i.e., the counting rate, increases, or as the pulse width, i.e., the time constant of the pulse shaper, increases.

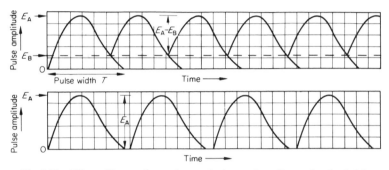

Fig. 4-39   Effect of incomplete pulse resolution on base-line and pulse height

A further complication occurs when a second pulse occurs effectively within the same time T as the first. These pulses are summed producing a pulse of height about $2 \times E_A$ with subsequent distortion of the high energy side of a given pulse amplitude distribution. This effect is called pulse pile-up and is generally minimized by use of a "pulse pile-up rejector". This device uses, in effect, a shorter time constant pulse shaper to interrogate the rate of incoming pulses. Where a second pulse arrives within the *normal* pulse width (i.e., time T), an artificial pulse is generated to replace the two coincident pulses. One consequence of the use of a pulse pile-up rejector is that the usual scheme of dead time correction outlined on p. 65 cannot be employed since a statistical distribution cannot be assumed. Alternative correction schemes have been proposed, however,[47] and hardware solutions can also be employed to compensate for dead time losses.

References

(1) Von Hevesey, G., *Chemical Analysis by X-rays*, McGraw-Hill, New York, 1932, Chapter 3.
(2) Coolidge, W. D., *Phys. Rev.* **2**, 409 (1913).
(3) Coster, D. and Nishina, J., *Chem. News* **130**, 149 (1925).
(4) Glocker, R. and Schreiber, H., *Ann. Physik* **85**, 1089 (1928).

(5) Coster, D. and Druyvesteyn, H., *Z. Physik* **40**, 756 (1927).

(6) Henke, B., *Advan. X-ray Anal.* **5**, 288 (1961).

(7) Sobolevskaya, G. D., Ioffe, Y. K. and Nikolaenko, G. M., *Zavodsk. Lab.* **31**, 1766 (1965).

(8) Walker, G., *Advan. X-ray Anal.* **9**, 504 (1966).

(9) Hans, A. *et. al.*, *Comptes Rendus du Colloque de Spectrométrie X, Bruxelles*, Philips, Eindhoven, 1964, p. 155.

(10) Wyckoff, R. W. G. and Davidson, F. D., *Rev. Sci. Instr.* **35**, 381 (1963).

(11) Campbell, W. and Thatcher, J. W., *Instrumentation for primary and secondary excitation of low energy X-ray spectral lines*, U.S. Bureau of Mines Report, R16689, 1965.

(12) Spielberg, N., Proceedings of the 3rd International Symposium on X-ray Optics and X-ray Microanalysis, Stanford University, 1962.

(13) Chadwick, J., *Phil. Mag.* **24**, 594 (1912).

(14) Jopson, R. A., Mark, H. and Swift, C. D., *Phys. Rev.* **127**, 1619 (1962).

(15) Birks, L. S., *X-ray Spectrochemical Analysis*, 2nd Edition, Wiley, New York, 1969.

(16) Duggan, J. L. *et al.*, *Advan. X-ray Anal.* **15**, 407 (1971).

(17) Johansson, T. B., Akselsson, R. and Johansson, S. A. E., *Advan. X-ray Anal.* **15**, 373 (1971).

(18) Hope, J. A. and Watt, J. S., *Int. J. Appl. Rad. Isotopes* **16**, 9 (1965).

(19) Watt, J. S., *Int. J. Appl. Rad. Isotopes* **18**, 383 (1967).

(20) Watt, J. S., *At. Energy (Australia)* **8(2)**, 6 (1965).

(21) Jenkins, R. and de Vries, J. L., *Practical X-ray Spectrometry*, 2nd Edition, Macmillan, London, 1972, Chapter 4.

(22) Curran, S. C. and Craggs, J. O., *Counting Tubes*, Butterworths, London, 1949.

(23) Curran, S. C., Angus, J. and Cockroft, A. L., *Phil. Mag.* **40**, 929 (1949).

(24) Fano, U., *Phys. Rev.* **70**, 44 (1946).

(25) Campbell, A. J. and Ledingham, K. W. O., *Brit. J. Appl. Phys.* **17**, 769 (1966).

(26) Jenkins, R., *Philips Scientific & Analytical Equipment Report, F.S. 6*, Philips, Eindhoven, 1970.

(27) Jenkins, R. and Hurley, P. W., *Can. Spectry.* **13**, 35 (1968).

(28) Elad, E., *Nucl. Instr. Methods* **37**, 327 (1965).

(29) Bowman, H. R. *et al.*, *Science* **151**, 562 (1966).

(30) Oosting, J. A., *An Introduction to semiconductor radiation detectors*, Philips A.I. Bulletin 440, Philips, Eindhoven, 1967.

(31) Gedke, D. A., *X-ray Spectrom.* **1**, 129 (1972).

(32) Jenkins, R. and de Vries, J. L., *Practical X-ray Spectrometry*, 2nd Edition, Macmillan, London, 1972, Chapter 2.

(33) Bertin, E. P., *Principles and Practice of X-ray Spectrometric Analysis*, Plenum, New York, 1970, Chapter 5.

(34) Liebhafsky, H. A., Pfeiffer, H. G., Winslow, E. H. and Zemany, P. D., *X-rays, Electrons, and Analytical Chemistry*, Wiley-Interscience, New York, 1972, Chapter 5.

(35) Jenkins, R., *X-ray Spectrom.* **1**, 23 (1972).

(36) Birks, L. S. and Seal, R. T., *J. Appl. Phys.* **28**, 541 (1957).

(37) Vierling, J., Gilfrich, J. V. and Birks, L. S., *Appl. Spectry.* **23**, 342 (1969).

(38) Jenkins, R., Croke, J. F., and Westberg, R. G. *X-ray Spectrom.* **1**, 59 (1972).

(39) Bertin, E. P., *Principles and Practice of X-ray Spectrometric Analysis*, Plenum, New York, 1970, p. 121.

(40) Jenkins, R. and de Vries, J. L., *Practical X-ray Spectrometry*, 2nd Edition, Macmillan, London, 1972, p. 38.

(41) *Energy Dispersion X-ray Analysis*, ASTM Special Technical Publication STP485, American Society for Testing and Materials, Philadelphia, 1970.

(42) Frankel, R. S. and Aitken, D. W., *Appl. Spectry.* **24**, 557 (1970).

(43) Russ, J. C., *Elemental X-ray Analysis of Materials*, EDAX Laboratories, Raleigh, North Carolina, 1972.

(44) Gammage, C. F., *X-ray Spectrom.* **1**, 99 (1972).

(45) Porter, D. E., *X-ray Spectrom.* **2**, 85 (1973).

(46) Goulding, F. S. and Stone, Y., *Science* **170**, 280 (1970).

(47) Landis, D. A., Goulding, F. S. and Jarrett, R. V., *Nucl. Instrum. Metals*, **101**, 127 (1972).

CHAPTER 5
qualitative analysis

5.1 QUALITATIVE ANALYSIS

X-Ray spectrometry is particularly well suited to qualitative analysis, since the technique is rapid and almost completely non-destructive, and measurable signals can be obtained from as little as a few milligrams of specimen. X-Ray spectra are relatively simple to interpret and in the case of crystal spectrometers there is little chance of gross interpretation errors where elemental concentrations are in excess of a few tenths of one percent. Below this concentration level there is a far greater chance of spectral interference from the higher concentration level elements but, even so, under favourable circumstances measurable signals can be obtained for elements at the ppm level. The chance of spectral overlap is far greater in the case of energy dispersion spectrometers, where, for the longer wavelengths, the energy resolution is worse than the crystal spectrometer. This effect was clearly demonstrated in Fig. 4-18 which indicates that even where the Si(Li) detector is used, the energy dispersive technique is one or two orders of magnitude worse than its crystal dispersion counterpart. The overall effect of this in qualitative (and indeed quantitative) analysis can be judged from Fig. 2-15 which indicates that the chance of overlap of K, L, and M spectra increases markedly at the long wavelength region. However, modern methods of spectral stripping can do much to enhance weak signals and overlapping spectra and this technique has found great use, more particularly in quantitative analysis. The situation in the short (<0.7 Å) wavelength region is to a large extent reversed and here the Si(Li) detector has better resolution than the crystal spectrometer.

Whichever technique of obtaining spectra is employed, the basic approach in data interpretation is similar. The spectrum is recorded ensuring that the required wavelength, or energy range is adequately covered, both from the point of view of excitation as well as detection. It has been shown in Section 2-9 (p. 21) that characteristic K spectra generally have similar line distributions for all atomic numbers, except for the very long wavelengths where satellite line and band spectra formation begin. Thus, if the presence of an element is suspected, because of the presumed identification of a Kα line, a corresponding (weaker) Kβ line *must* also be present. If the Kβ line cannot be confirmed because of an interfering line, it is probable that a second or third order K series line could be used for confirmation of the presence of the element. Attempts to identify an element by means of one line alone are dangerous, particularly in the short wavelength region where K spectra are crowded. A similar situation is true for the L spectra where the suspected presence of an Lα line *must* mean that Lβ's and Lγ's are also present.

5.1.1   Qualitative and semi-quantitative analysis with the crystal spectrometer

The data output on the crystal spectrometer is generally in the form of a chart recording of °2θ versus intensity and standard tables are available[1] for the conversion of °2θ to wavelength and/or atomic number.

Figure 5-1 shows a typical spectrum from a specimen of mixed selenium, zirconium, and antimony oxides obtained with a crystal spectrometer. The Kα_{1,2}, Kβ_{1,3}, Kβ_2 series of lines is clearly visible for the three elements in question, the intensity ratios in each case being about 100:15:2. This particular specimen also contains a certain amount of iodine and again the characteristic K spectrum is clearly visible in the 10–13 °2θ region. It will be seen that both antimony and iodine give second order lines within the spectral region covered, in each case the second order lines being about 35% as strong as the first order line. This dropping off in intensity for higher order reflection is also illustrated

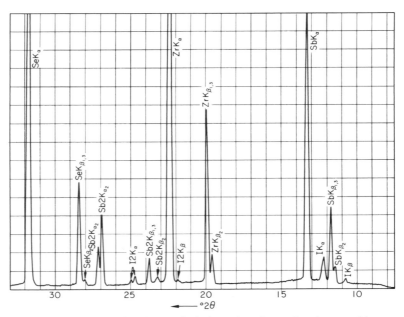

Fig. 5-1   Spectrum of a mixture of selenium, zirconium and antimony oxides

in Fig. 4-26 and the general rule of thumb is that intensities drop by a factor of about 3 for each successive increase in the order of reflection.

It is frequently possible to obtain useful semi-quantitative analyses by comparing the relative intensities of similar lines in a chart record. For instance, in Fig. 5-1 the relative intensities of Se, Zr, Sb, and I, $K\beta_{1,3}$ lines are $28:48:20:2$. Since the mixture is stated to be a mixture of oxides, these intensities have to be scaled by the appropriate factors to convert element to oxide, before the data can be normalized to 100%. Table 5-1 shows the various steps in the calculation and the estimated composition. This particular specimen actually contained about 50% of ZrO_2 and 25% each of SeO_2 and Sb_2O_5. Zirconium iodate was a known impurity.

TABLE 5-1.
Estimated composition of oxide mixture

Element	$K\beta_{1,3}$ rel. intensity	Oxide factor	Estimated composition
Se	28	$1.16 \times 28 = 32.5$	27% SeO_2
Zr	48	$1.23 \times 48 = 59.0$	49% ZrO_2
Sb	20	$1.35 \times 20 = 27.0$	22% Sb_2O_5
I	2	$1.24 \times 2 = 2.5$	2% IO_3'

Figure 5-2 shows the spectrum obtained from a specimen of solder which contains tin and lead as its major constituents. This spectrum differs from that dealt with previously, since in this instance both K and L series lines are present. The additional complexity of the pattern can be seen by comparing the first order K lines of tin, at around $12-14\,°2\theta$, with the L lines of lead, which occur over the $22-35\,°2\theta$ range. Since the L spectrum contains three times as many lines as the K spectrum in the former case the chance of spectral overlap is greater. A typical example is the overlap of the Pb $L\beta_{1,2}$ doublet with the second order Sn $K\alpha_{1,2}$ doublet at $28°2\theta$. The specimen also contains traces of antimony and bismuth. In the previous specimen it was possible to estimate the approximate composition by directly comparing the heights (i.e., intensities) of certain lines. This is more difficult in the present example because firstly, it is not possible to compare intensities of *similar* lines from different elements, and secondly, because there is quite a difference in the excitation potentials of the elements present. Nevertheless, an estimation of composition can be made by allowing for these two factors. A correction can be applied for the different line series by use of fluorescent yield values weighted for the lines in question. The effect of difference in excitation potentials can be corrected for using equation (6-1). It will be seen that characteristic line intensity is proportional to $(V - V_c)^{1.6}$ where V is the X-ray tube potential (50 kV in this instance) and V_c the critical excitation potential (13 kV for lead and bismuth, 29 kV for tin and 30 kV for

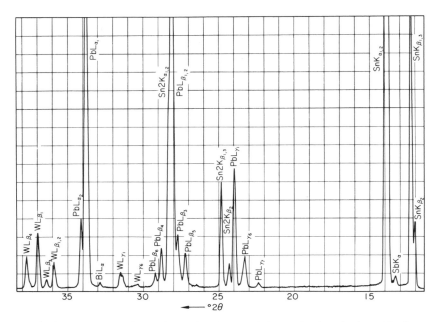

Fig. 5-2   Spectrum of a specimen of solder containing predominantly tin and lead

antimony). Table 5-2 shows the various steps in the calculation and the estimated composition. The given values are in reasonable semi-quantitative
agreement with the stated composition of 50/50 lead, tin with traces of bismuth
and antimony.

   Figure 5-2 also shows the presence of tungsten L lines in the 30–38°2θ region.
These lines arise from scatter from the tungsten anode X-ray tube. An X-ray
spectrum always contains characteristic lines from the X-ray tube anode, and
these may prove a nuisance—both from the point of view of spectral interference,
as well as complicating the identification of a sample containing the same element as the X-ray tube anode. The intensities of the tube lines will vary with the

TABLE 5-2.
Estimated composition of lead/tin mixture

Element	Line	Intensity I_0	Weighted fluorescent yield/ω	Excitation factor F	Corrected intensity $I_c = I_0/\omega F$	Estimated concentration /%
Pb	Lα	180	0.2	$[50-13]^{1.6} = 323$	2.79	53.2
Bi	Lα	1.5	0.2	$[50-13]^{1.6} = 323$	0.02	0.4
Sn	Kα	250	0.8	$[50-29]^{1.6} = 130$	2.40	45.8
Sb	Kα	2.5	0.8	$[50-30]^{1.6} = 121$	0.03	0.6

scattering power of the specimen and will be most intense for low average atomic matrices.

5.1.2 Qualitative and semi-quantitative analysis with the energy dispersion spectrometer

The output device for qualitative work on the energy dispersion spectrometer is generally a video display system in which intensity is shown as a function of photon energy. As in the case of the crystal spectrometer, tables are available for the direct conversion of photon energy to atomic number.[2] It is now becoming common practice to incorporate a small digital computer into the data handling system of an energy dispersion spectrometer, which can be used for peak stripping, conversion of X-ray intensity to concentration and, in this instance, for the conversion of photon energy to atomic numbers. Where these facilities are available, it is possible to have automatic identification of lines, with the atomic number and/or line transition displayed along with the spectrum.

The power of the energy dispersion spectrometer lies mainly in the speed at which data may be recorded. Figure 5-3 shows the spectrum of a solder specimen[3] similar to that obtained with the crystal spectrometer. Although the resolution obtained with the energy dispersion is much worse than that obtained with the crystal spectrometer, the energy dispersion spectrum was recorded in about one tenth the time required for the crystal spectrometer. Even though in this instance the energy dispersion data are worse, in terms of absolute resolution, the data are more than sufficient to demonstrate the presence and approximate relative concentration levels of the major elements.

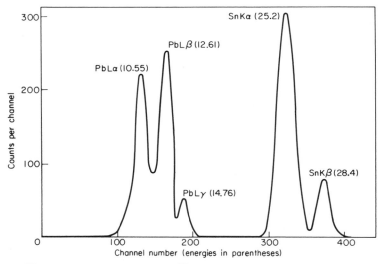

Fig. 5-3   Analysis of solder with ^{57}Co radiation source and Si (Li) detector

As stated previously, the resolution of the energy dispersion spectrometer is better for the shorter wavelengths than for longer wavelengths. Figure 5-4 gives an idea of how good this resolution can be for the very short wavelengths and shows the K spectra for elements in the atomic number region 82(Pb)–92(U). Lead Kα has a wavelength of 0.167 Å and uranium Kα 0.128 Å.

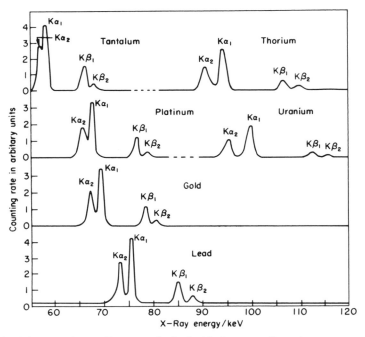

Fig. 5-4   Pure element K spectra obtained with the energy dispersion spectrometer

Modern energy dispersion spectrometers generally have a miniature X-ray tube as the excitation source and this gives similar broad wavelength range excitation as in the case of the crystal dispersion system. A further recent development has been the use of a source consisting of a high power X-ray tube in combination with a suitable secondary fluorescer. Careful choice of this secondary target gives a selectable monochromatic source which in turn may offer specific excitation for the portion of the wavelength range of interest. Figure 5-5 shows the spectrum obtained from the National Bureau of Standards Orchard Leaf Standard, which contains such elements as Na, Mn, Ni, Cu, Zn, As, Rb and Pb at concentration levels of the order of a few tens of micrograms per gram. The spectrum was obtained[4] in a counting time of 1000 seconds, using a tungsten target tube with a molybdenum secondary target.

Semi-qualitative information can be obtained directly from the recorded spectra, using a similar type of approach as in the crystal dispersion system.

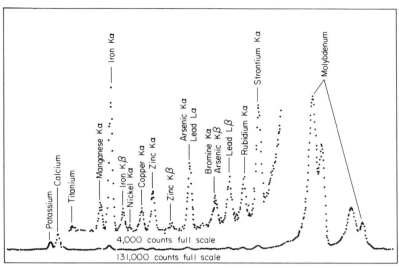

Fig. 5-5 Spectrum of N.B.S. Standard Orchard Leaves, recorded with the energy dispersion spectrometer;[4] specimen: plant leaf; time: 1000s; source: tungsten tube; secondary target: molybdenum.

Provisional analysis for orchard leaf* N.B.S. standard 1571.
Trace constituents in μg/g

Iron	300 ± 20	Copper	12 ± 1
Manganese	91 ± 4	Rubidium	12 ± 1
Sodium	82 ± 6	Nickel	1.3 ± 0.2
Lead	45 ± 3	Mercury	0.155 ± 0.015
Boron	33 ± 3	Cadmium	0.11 ± 0.02
Zinc	25 ± 3	Selenium	0.08 ± 0.01
Arsenic	14 ± 2	Uranium	0.029 ± 0.005

Major constituents in wt. %		Minor constituents in wt. %	
Nitrogen	2.76 ± 0.05	Magnesium	0.62 ± 0.02
Calcium	2.09 ± 0.03	Phosphorus	0.21 ± 0.01
Potassium	1.47 ± 0.03		

The same precautions must, however, be observed particularly from the viewpoints of excitation and detection efficiency. A further point to be remembered in the case of energy dispersion data, is that it is the *area* rather than the height of the peak that is a measure of the characteristic line intensity. Equation (4-11) shows that the resolution decreases with decrease of energy, (and hence with the absolute resolution of the detector) thus the Kα line from, for example, potassium, will be much broader than the Kα line from tin. This is not true in the case of the crystal dispersion spectrometer where the line width is determined by the collimation of the spectrometer and mosaic spread of the analysing crystal. In this instance, for a given collimator/crystal combination, peak height is always proportional to peak area.

References

(1) White, E. W. and Johnson, G. G. Jr., *X-ray and Absorption Wavelengths and Two-Theta Tables*, ASTM Data Series DS37A, American Society for Testing and Materials, Philadelphia, 1970.

(2) *X-ray Emission Wavelengths and KeV Tables for Non-Diffractive Analysis*, ASTM Data Series DS46, American Society for Testing and Materials, Philadelphia, 1971.

(3) Bowman, H. R., Hyde, E. K., Thompson, S. G. and Jared, R. C., *Science* **151**, 562 (1966).

(4) Woldseth, R., *X-ray Energy Spectrometry*, Kevex, Burlingame, 1973, 3.12.

errors in X-ray analysis

6.1 TYPES OF ERROR IN X-RAY ANALYSIS

Like most methods of instrumental analysis, X-ray spectrometry is a comparative technique and the accuracy of a measured result will depend not only upon the precision of the instrumental data but also upon the accuracy of the calibration standards. It is important to establish the types and sources of the various errors in order that these can be brought within acceptable proportions. Measurements of this type are generally subject to three types of errors, random errors, systematic errors and wild errors. Figure 6-1 illustrates the essential differences between random and systematic errors. If a measurement is repeated many times and a graph made of frequency of result versus result, a distribution similar to that shown in Fig. 6-1(a) may be obtained. Here the distribution of results is symmetrical about the "true" result and the arithmetic mean of all results is equal to the "true" result. Where this situation occurs, the error involved is a random error and the magnitude of this error can be estimated by normal statistical procedures.[1] It should be appreciated that, in practice, the "true" result may not be known and some degree of uncertainty is generally associated with its value. Figure 6-1(b) illustrates the case of a combined systematic and random error. As before, a measurement is repeated many times and the distribution of these results is found to be symmetrical about the arithmetic mean. Again this demonstrates that a random error is associated with a single measurement, but in this instance, the arithmetic mean does not correspond to the "true" result. This indicates that a systematic error is also present and the magnitude of this error is given by the absolute difference between the arithmetic mean and the "true" result. The precision of a measurement is dependent only upon the random errors whereas the accuracy of a measurement depends on both random *and* systematic errors. Thus a result can be precise but inaccurate and this also is illustrated in Fig. 6-1.

In Fig. 6-1(a) the indicated point has an associated uncertainty indicated by the error limits, and the precision of the measurement is $0.5 \pm 0.1\%$. The same precision would be reported for the indicated point in Fig. 6-1(b), that is $\pm 0.1\%$,

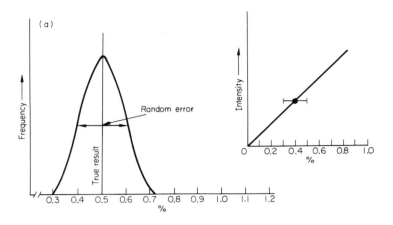

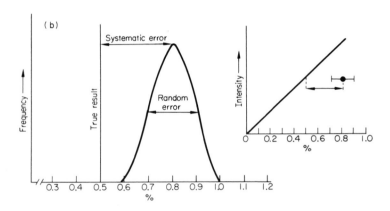

Fig. 6-1   Effect of random and systematic errors on precision and accuracy

but in this instance the measured result is 0.8 %. Thus, in this example, the result is inaccurate by 0.3 %.

A point to be remembered in X-ray spectrometry is that the error limits are generally controlled in terms of X-ray intensity rather than in terms of concentration. These will be the same where a straight line calibration is used and where this line passes through the point $I_x = 0$; $C_x = 0$. Figure 6-2 shows that where the calibration curve is flattened, or where a large intercept (that is a high background) is present, a given error in the intensity may well correspond to a greater error in the concentration. Where the calibration curve is very flattened, i.e., a small change in intensity corresponds to a large difference in concentration, the method is said to lack sensitivity.

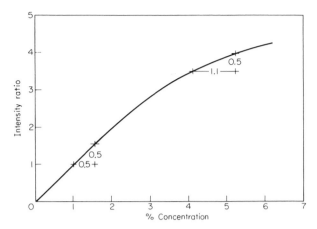

Fig. 6-2   Loss of sensitivity due to flattening of the calibration curve

6.2   SOURCES OF RANDOM ERRORS IN QUANTITATIVE X-RAY SPECTROMETRY

The major sources of random errors in an X-ray measurement are those due to the equipment and to the random nature of X-ray production, the latter being generally referred to as "counting statistics". Typical of an instrumental random error is that due to the source, which usually consists of an X-ray tube and high voltage generator. The function of the generator is to provide a stabilized high voltage and current to the X-ray tube and in practice it is found that for a single element, the measured intensity of a given wavelength $I(\lambda)$, excited by a white radiation source, is related to voltage V and current i in the following way:

$$I(\lambda) = Ki(V - V_c)^{1.6} \qquad (6-1)$$

where K is a constant which depends upon the anode material of the X-ray tube and the instrumental conditions employed, and V_c is the critical excitation potential of the wavelength λ. For optimum excitation conditions V is usually set to three to five times V_c. In the simplest case, that is where there are no significant systematic errors present, there will be a linear relationship between the concentration of the element from whence λ was excited, and $I(\lambda)$. It will be seen from equation (6-1) that should the current on the X-ray tube vary, for example by 1 %, during the measurement cycle, there will be an error of 1 % in $I(\lambda)$, with a subsequent error of 1 % in the calculated concentration. Similarly, if the potential on the X-ray tube should vary by 1 % during the measurement cycle, the variation in $I(\lambda)$ will be more than 1 %—depending upon the relative values of V and V_c, with a corresponding error in the concentration. It is thus clear that if the concentration is to be determined to an accuracy of 0.1 %, then both current and potential have to be stabilized to at least twice as good as this value. X-Ray generators designed for use with X-ray spectrometers are generally stabilized to about 0.05%.

Other random errors arising from the equipment include those due to the mechanical resetting of instrumental parameters such as goniometer, analysing crystal, detector, voltage, pulse height selector and so on. In a typical system these errors are rarely in excess of 0.1%.

The error due to counting statistics is a truly random error and is found to obey the normal statistical rules for handling random events. Since the number of events—in practice the number of counts N—is always large (generally at least 10^4), the distribution of data is found to follow a Gaussian distribution. This allows us to predict that the standard deviation σ of a given number of collected counts N is given by:

$$\sigma = N^{1/2} \qquad (6\text{-}2)$$

or in other words 68.3% of an infinite series of measurements of N, having an arithmetic mean of $\overline{N}$, would lie between the limits $\overline{N} \pm \overline{N}^{1/2}$. Similarly 92% ($2\sigma$) would lie between the limits $\overline{N} \pm 2\overline{N}^{1/2}$, and 99% ($3\sigma$) between $\overline{N} \pm 3N^{1/2}$. In general the analyst has to be content with a single measurement, or at the most two or three, and again we are able to use the properties of the Gaussian distribution to predict the likely error in the determination. For example, if under a given set of conditions, 10^6 counts were collected, we are 68.3% certain that the true value of the number of counts lies between $10^6 \pm (10^3)$. Similarly we are 92% certain that the value lies between $10^6 \pm (2 \times 10^3)$, and 99% certain that the value lies between $10^6 \pm (3 \times 10^3)$. Most analytical spectroscopists work with the 92%, or 2σ, confidence limits.

It is often more useful to work with the percentage standard deviation or

$$\sigma\% = N^{1/2} \times \frac{100}{N} = \frac{100}{N^{1/2}} \qquad (6\text{-}3)$$

and since the number of counts taken is the product of the counting rate I and the counting time T, it follows that the percentage standard deviation is also given by:

$$\sigma\% = \frac{100}{(IT)^{1/2}} \qquad (6\text{-}4)$$

It will be noted that equation (6-4) relates simply to the total counting rate I, whereas, in practice, one is generally faced with the problem of estimating the error σ_{net} in the net counting rate of peak I_p and background I_b counting rates. It can be shown[2] that the general form of the relationship between σ_{net}, I_p and I_b is:

$$(\sigma\%)_{net} = \frac{100}{T^{1/2}} \frac{1}{I_p^{1/2} - I_b^{1/2}} \qquad (6\text{-}5)$$

where T is the total time spent counting on the peak T_p, plus that spent on the background T_b. One of the conditions for equation (6-5) is that the total time should be distributed between peak and background in the following way:

$$T_p/T_b = (I_p/I_b)^{1/2} \tag{6-6}$$

This method is generally referred to as the method of "optimum fixed time". As an example, if the counting rates on peak and background were 2500 counts/s and 900 counts/s respectively, the total time required to achieve a net counting error of 0.5% (σ) would be:

$$T^{1/2} = \frac{100}{0.5} \frac{1}{(2500)^{1/2} - (900)^{1/2}} \quad \text{or} \quad T = 100 \text{ seconds}$$

This 100 seconds should be divided up such that $T_p/T_b = 2500/900$ or $T_p = 62$ seconds and $T_b = 38$ seconds.

It will be seen from equation (6-5), that where the value of the peak counting rate is large, relative to the background, the background can be ignored. For all practical purposes this situation occurs where I_p/I_b is greater than ten, or in other words, the background can be ignored where the peak to background is greater than $10:1$.

Another consequence of equation (6-5) is that for a fixed analysis time, the net counting error will be a minimum when $I_p^{1/2} - I_b^{1/2}$ is as large as possible. For this reason the numerical difference between the square roots of peak and background counting rates is often taken as a "figure of merit", for the establishment of optimum experimental conditions.

6.3 THE LOWER LIMIT OF DETECTION

It will be seen from equation (6-5) that as $I_p \to I_b$, $\sigma_{net} \to \infty$. In other words, as the peak counting rate approaches the same value as that of the background, the net counting error becomes infinite. This introduces the concept of the lower limit of detection which is often defined as "that concentration equivalent to two standard deviations of the *number* of background counts is equal to 2σ or N_b. Since counting rate is number of counts divided by time, if we can assume no significant error in the measurement of T (generally true) then

$$2\sigma_{(I_b)} = \frac{2\sigma_{(N_b)}}{T_b} = \frac{2(N_b^{1/2})}{T_b} = \frac{2(I_b T_b)^{1/2}}{T_b} = 2\left(\frac{I_b}{T_b}\right)^{1/2}$$

To convert intensity to concentration it is necessary to divide by m, the counts/s per $\%$ for the element in question, thus:

$$2\sigma_{(I_b)} \text{ (in terms of concentration)} = \frac{2}{m}\left(\frac{I_b}{T_b}\right)^{1/2}$$

In practice it is necessary to perform two measurements in order to determine a "low" concentration, i.e., peak and background or peak and blank. This in turn means that the error will be increased by the square root of two, thus

$$\text{Lower limit of detection} = \frac{2(2)^{1/2}}{m}\left(\frac{I_b}{T_b}\right)^{1/2} \simeq \frac{3}{m}\left(\frac{I_b}{T_b}\right)^{1/2} \qquad (6\text{-}7)$$

The expression given in equation (6-7) is that generally employed to estimate the detection limit of a given element measured under a specified set of conditions. It should be noted that T_b is the time spent counting on the background and since, at the detection limit, $I_p \simeq I_b$, it follows from equation (6-6) that $T_p = T_b$, thus T_b in equation (6-7) is *one half* of the available counting time. As an example, if a given element at the 0.2% concentration level, gave a peak counting rate of 330 counts/s and a background of 30 counts/s, m would equal 1500 counts/s per %, and the detection limit in 120 s would be about 0.0014% or 14 ppm.

It does not follow, however, that one could in practice, determine 14 ppm of the element since, as has already been pointed out, at the detection limit the error is infinite. There is thus a subtle difference between the lower limit of *detection* and the lower limit of *determination*. A good rule of thumb is that the lower limit of determination is about three times the lower limit of detection, or in other words, a concentration equivalent to six standard deviations of the background counting rate. For instance, in the case cited above, the lower limit of determination would be equivalent to about 42 ppm. The actual error can be easily estimated since we know that the element gives 1500 counts/s per %, so 42 ppm would give 6.3 counts/s or an I_p of $(6.3 \pm I_b) = 26.3$ counts/s. Substitution in equation (6-5) gives an error of about 17% or 7 ppm.

Table 6-1 lists a few typical sensitivity and detection limit data for some of the lower atomic number elements. It will be seen that detection limits vary not

TABLE 6-1.
Detection limits for low atomic number elements

Element	Matrix	Counts/s per %	Background (counts/s)	Lower limit of detection/ppm†
Na	Al_2O_3	108	42	253
Mg	Limestone	646	60	51
Mg	Al	1,360	470	67
Si	Steel	2,300	105	19
Si	Limestone	2,950	90	14
P	Oil	24,000	450	4
P	Nylon	10,000	13	15
S	Oil	63,000	170	0.8

† Lower limit of detection calculated for the 2σ confidence level with 100 s counting time.

only with the atomic number of the measured element but also with the sample matrix. This is because both m and I_b are affected by changes in the matrix. In general, I_b increases with decrease in the average atomic number of the matrix, and m decreases with increase in the total absorption of the matrix for the measured wavelength.

The detection limits are generally in the low part per million levels for most of the atomic number range, although there is a decrease in the sensitivity at the wavelength extremes of the range of the spectrometer. Figure 6-3 illustrates the reasons for this fall-off in sensitivity and shows typical calibration curves for three selected portions of the wavelength region.[3] The short wavelength region (0.2–0.8 Å) is typified by reasonably large values of counts/s per % but also by high backgrounds. The counts/s per % achievable are not as good as might be expected since, in this region, it is impossible to obtain a potential

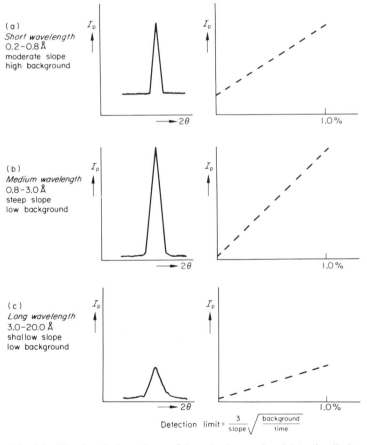

$$\text{Detection limit} = \frac{3}{\text{slope}}\sqrt{\frac{\text{background}}{\text{time}}}$$

Fig. 6-3   Wavelength dependence of slope, background and detection limit

on the X-ray tube much in excess of the critical excitation potential. As will be seen from equation (6-1), this means that the excitation conditions are far from optimum. Background values are high in this region due mainly to the inefficiency of the pulse height selector. The majority of the background arises from first order scattered continuum, and, as was explained in Section 4.2.4 (p. 73), to achieve efficient use of the pulse height selector, energy differences of at least 50 % are required. The medium wavelength range (0.8–3 Å), is typified by high values of counts/s per % and low background. In this instance there is no problem in obtaining optimum excitation and the pulse height selector is able to remove most of the scattered background. The same is essentially true of the long wavelength region (3–20 Å) but here absorption problems are particularly severe, reducing the long wavelength contribution from the X-ray tube, hence causing poorer excitation and also causing significant attenuation of the characteristic wavelengths by the detector window.

6.4 SYSTEMATIC ERRORS IN QUANTITATIVE X-RAY SPECTROMETRY

Probably the greatest source of both systematic and wild results is the operator himself. This is especially true in the case of so-called manual spectrometers, in which the operator has to take the responsibility for the selection of a variety of parameters such as the correct 2θ value or energy range, the correct excitation conditions and so on. In completely automatic spectrometers the task of the operator is limited to the correct insertion and identification of the specimen and, in this case, the chance of error is greatly reduced. A second source of systematic error is the equipment, where such effects as dead time, pulse shift and detector instability, can all cause significant systematic errors to accrue, unless some form of automatic or manual compensation is applied. In the majority of automatic spectrometers such provision is made and the systematic errors are again generally less than 0.1 %. In less sophisticated systems, however, the responsibility lies with the operator and a common technique employed in this circumstance is the so-called method of ratio counting, whereby a reference standard is measured under conditions identical to those under which the sample is analysed, care being taken to complete both sample and reference cycles within a relatively short time interval, thus reducing any additional problems due to long term "drifts" in the equipment.

A common source of instrumental error is when nearby wavelengths overlap. This situation can arise where insufficient resolution has been employed. In cases where sufficient resolution just is not available, a correction for line overlap has to be included in the intensity/concentration algorithm.

Assuming for the moment a "perfect" spectrometer operated by a "perfect" spectroscopist, the last and most important source of systematic errors is the sample.

6.5 SAMPLE PREPARATION

In addition to the obvious problems of mutual interference due to the absorption and enhancement effects already discussed in Section 3-2 (p. 40), the preparation of the specimen to be analysed can be a critical factor where high accuracy is required. Since X-ray spectrometry, like most instrumental methods of analysis, is essentially a comparative technique, it is absolutely vital that calibration standards and specimens to be analysed are presented to the spectrometer in a reproducible and identical manner. Since the effective penetration depth of characteristic radiation is relatively short—perhaps 1 to 100 μm—it is important that the thin layer of the sample actually contributing to the measured characteristic wavelength is both homogeneous and representative of the bulk sample. Although the irradiated area is quite large, of the order of 7 cm^2, due to the small penetration, the actual volume analysed lies between 0.01 and 0.1 cm^3. Generally, surface heterogeneity effects are more critical for the longer wavelengths, thus more important in the case of the lower atomic numbers, so particular care is required in the preparation of samples containing such elements as F, Na, Mg, Al and Si. Whichever form of sample preparation is employed, speed is nearly always of paramount importance, particularly today, where multi-channel spectrometers can analyse specimens, and process the data, at the rate of more than one a minute. The actual sample handling procedure eventually chosen is almost invariably a compromise between accuracy and speed.

Bulk metal specimens, such as discs from castings, need only be surfaced by milling or surface grinding and this can generally be carried out in a matter of minutes. Liquids also present no great difficulty, provided that they are stable under the influence of the X-ray beam and provided that the X-ray spectrometer is fitted with a helium flushing system, thus allowing the measurement of the longer wavelengths without a vacuum.

The greatest difficulties are undoubtedly encountered in the analysis of powder specimens where local heterogeneity within particles tend to complicate specific absorption effects. These problems are classified under the broad heading of "particle size effects" but it is useful to make some further categorization in order that the problems be better understood. Figure 6-4 illustrates a somewhat idealised situation in which a two phase specimen is shown to be made up of large particles of phase 1 and small particles of phase 2. Let us assume that phase 1 is chalkopyrite $CuFeS_2$ and phase 2 is pyrites FeS_2. If the element being measured is present only in phase 1, for example copper, and if the matrix absorption coefficients are the same for both phases, where the effective penetration of the measured wavelength is small in comparison with the average size of the particles in phase 1, a *grain size effect* would result. This will be apparent from the figure since it will be seen that there is a higher concentration of phase 2 at the surface of the sample than in the bulk. Reducing the average particle size of phase 1 causes the concentration of the measured element in the analysed layer to

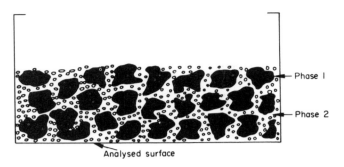

Fig. 6-4   The particle effect in X-ray fluorescence spectrometry

	Fluorescing element	Absorption for Measured wavelength	Effect
Phase 1 Phase 2	present absent	same	grain size
Phase 1 Phase 2	present absent	different	inter-mineral
Phase 1 Phase 2	present present	different	mineralogical

increase, with a subsequent increase in the characteristic line intensity. This situation is further complicated where the absorption coefficients of the two phases for the measured wavelength are different and this effect is classified as an *inter-mineral effect*. The third and most complicated situation arises where the measured element is present in both phases, for example, iron in the case cited previously, and where each phase has a different absorption coefficient for the measured wavelength. This third case is by far the most common in practice, and is classified as a *mineralogical effect*.

Each of these three effects can be minimized by reducing the particle size using a grinding device. The effects cease to be of importance when the particle diameter has been reduced to about one fifth of the penetration depth of the measured wavelength. However, as has already been shown, the penetration depth might be as small as a few microns and this in turn would require a maximum particle diameter of less than one micron. Particle sizes of this order are very difficult to achieve and there is obviously a practical limit as to what can be achieved within a reasonable grinding time. For the average disc-type mill, average particle sizes of 1 to 10 microns can be achieved for most materials in a matter of a minute or two. This may prove sufficient, even for the very long wavelengths, provided great care is taken in reproducing the grinding and subsequent pelletising procedure for both calibration standards and samples to be analysed. In perhaps five to ten percent of cases this technique does not give sufficient accuracy and in these instances recourse has to be made to a much more severe sample preparation procedure such as fusion to a glass. Many different types of fluxing materials have been employed but those based on sodium or lithium tetraborate are by far the most common. The flux can be made more acidic or more basic by addition of the appropriate fluoride or carbonate.

The successful preparation of fused discs depends very much upon getting the fusion reaction to go to completion and to avoid the formation of crystalline reaction products. Norrish and Hutton[4] have proposed the use of a mixture of lithium tetraborate and lithium carbonate, with lanthanum oxide added as a heavy absorber, for the preparation of whole rock samples. This technique has found general application since it offers several advantages. In the first case, addition of the lithium carbonate to the tetraborate forms an eutectic mixture which has a melting point several hundred degrees lower than that of the pure tetraborate. Second, the use of the lanthanum oxide does much to stabilize the matrix in that it has a high mass absorption coefficient for most of the rock forming elements, hence where standards and unknowns are prepared in the same way, the range of absorption coefficients is greatly reduced. One drawback with this technique, however, is the loss in sensitivity particularly for the lower atomic number elements. Dilution of a limited quantity of specimen may also be useful in that it gives a more manageable volume of sample. The major disadvantage with fusion methods is that they are time consuming. On the other hand, automatic fusion devices are now becoming available and these allow a fairly high throughput of specimens.

Figure 6-5 shows a very generalized scheme illustrating the more common steps involved in an X-ray analysis. Several methods of sample preparation are indicated and the dotted line indicates a typical analysis scheme which might have been followed in the analysis of the chalkopyrite/pyrites mixture referred to earlier.

6.6 ESTIMATION OF ERRORS IN QUANTITATIVE ANALYSIS

A very necessary part of any analytical procedure is the estimation of the accuracy of the working curve. The standard deviation s of a number of measurements n, involving a single data point x, can be estimated from the relationship:

$$s = \left(\frac{\sum (x - \bar{x})^2}{n - 1} \right)^{1/2} \tag{6-8}$$

where $\bar{x}$ is the arithmetic mean of the measurements. Note the use of s rather than σ for standard deviation in this instance. Strictly speaking, σ should be reserved for an infinite series of measurements, on a single data point, having

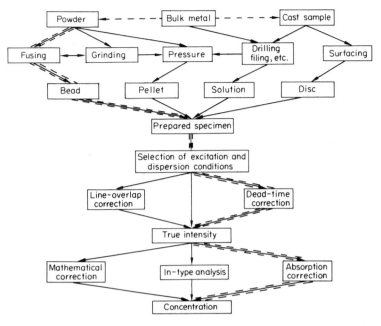

Fig. 6-5   Summary of the common steps utilised in X-ray fluorescence analysis

a true value of t. The following relationship then holds:

$$\sigma = \left(\frac{\sum (x - t)^2}{n}\right)^{1/2} \qquad (6\text{-}9)$$

A useful way of estimating the standard deviation over a range of concentration C is to use the relationship:

$$s = KC^{1/2} \qquad (6\text{-}10)$$

where K is a calibration constant. Figure 6-6 illustrates the use of this technique for estimating the value of K for a series of measurements of chromium in alloyed steel specimens. A plot has been made of the numerical difference between chemical and X-ray result against chemical concentration and limit lines have been drawn representing a K value of 0.02 for $1s$ and $2s$ confidence limits. If the chosen value of 0.02 for K is correct, 68 % of the data points should lie between the $1s$ limits, 92 % between $2s$ limits, and so on. In practice, the value of K is determined by adjusting the limit lines to give the correct distribution of points between the $1s$ and $2s$ confidence limits.

Typical values of K lie between 0.005 and 0.05. The former value could be called "good" analytical accuracy and the latter "poor" analytical accuracy.

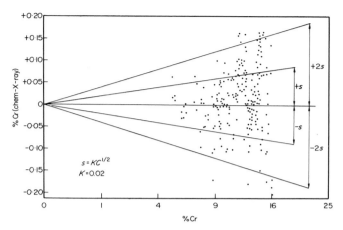

Fig. 6-6 Estimation of the average error over a range of concentration

References

(1) Beers, Y., *Introduction to the Theory of Error*, Addison-Wesley, New York, 1953.
(2) Jenkins, R. and de Vries, J. L., *Practical X-ray Spectrometry*, 2nd Edn., Macmillan, London, 1969, p. 188.
(3) Jenkins, R. and de Vries, J. L., *Analyst* **94**, 447 (1969).
(4) Norrish, K. and Hutton, J. T., *Geochim. Cosmochim. Acta* **33**, 431 (1969).

quantitative analysis

All methods of quantitative X-ray spectrometry involve three stages—the excitation of characteristic wavelengths from a suitably prepared specimen, the separation of the polychromatic beam of radiation allowing the intensities of individual wavelengths to be measured, and the correlation of the measured intensities with chemical composition. The calibration of a spectrometer for repetitive analysis of a given type of material generally involves the selection of suitable excitation and measuring conditions, followed by the derivation of a suitable algorithm relating measured intensity and elemental concentration. Also included would be the establishment of a suitable specimen preparation technique to ensure that the limited volume of specimen that contributes to the X-ray emission is really representative of the whole. A further complication which must always be considered is the effect of the sample absorption on the intensity of the measured wavelength. This gives rise to so-called matrix effects which may have to be minimized if accurate data is required.

The following sections discuss the relationship between X-ray intensity and the sample matrix and describes the more common methods of minimizing matrix effects.

7.1 BASIS OF QUANTITATIVE ANALYSIS

In Section 3-2 (p. 40) it was intimated that a complex relationship exists between the intensity $I(\lambda_i)$ of a wavelength arising from an element i in a matrix made up of elements j. It was shown that in homogeneous specimens three types of excitation and two types of absorption are involved.

(a) Direct excitation of λ_i by the primary source of radiation.

(b) Enhancement of λ_i by a matrix element A.

(c) A third element effect where a matrix element B enhances A, which in turn enhances i.

(d) Primary absorption, leading to modification of the primary radiation by matrix elements other than i.

(e) Secondary absorption, where all matrix elements (including i) absorb λ_i.

Of these five phenomena, (a) and (d) are by far the most important. The other three may or may not be significant, depending upon the matrix elements and the excitation conditions employed. In many instances, the analyst is probably unaware of the relative importance of the listed effects and where this is the case it is important to utilize an algorithm, relating intensity with concentration, which incorporates all of the five phenomena. Only where the matrix is well established can short cuts safely be taken, and simplified algorithms used.

7.2 RELATIONSHIP BETWEEN X-RAY INTENSITY AND THE SAMPLE MATRIX

The expression already given in equation (3-9) includes terms for direct excitation of λ_i by the tube spectrum, plus terms for primary and secondary absorption. It does not, however, include items for enhancement or third element effects. As long ago as 1955, Sherman[1,2] proposed relationships including absorption and enhancement effects, but more recent Japanese work[3,4] has presented algorithms in more practical forms. The expression given in Table 7-1 is due to Shiraiwa and Fujino[4] and contains terms for primary and secondary (enhancement) fluorescent effects. Their original work also contains a tertiary (third element) fluorescence term but since for most work the effect of this term is generally considered insignificant, it has been left out here for the sake of simplicity.

TABLE 7-1.
Primary and secondary fluorescence

$$I_p(\lambda_i) = \frac{1}{\sin \psi_2} \cdot \int_{\lambda_{min}}^{\lambda_{edge}^i} J(\lambda) \frac{Q_i(\lambda)}{\mu_s(\lambda) \sin \psi_1 + \mu_s(\lambda_i) \sin \psi_2} \, d\lambda$$

$$I_s(\lambda_i) = \frac{1}{2 \sin \psi_2} \sum_{j=1}^{n} \int_{\lambda_{min}}^{\lambda_{edge}^j} J(\lambda) \frac{Q_j(\lambda)Q_i(\lambda)}{\mu_s(\lambda) \sin \psi_1 + \mu_s(\lambda_i) \sin \psi_2} \left\{ \frac{\sin \psi_2}{\mu_s(\lambda_i)} \ln \left[\frac{1 + \mu_s(\lambda_i) \operatorname{cosec} \psi_2}{\mu_s(\lambda_j)} \right] \right.$$

$$\left. + \frac{\sin \psi_1}{\mu_s(\lambda)} \ln \left[\frac{1 + \mu_s(\lambda) \operatorname{cosec} \psi_1}{\mu_s(\lambda_j)} \right] \right\} d\lambda$$

$$Q_i(\lambda) = \mu_i(\lambda)W_i \left[\frac{r_i - 1}{r_i} \right] \omega_i g_i$$

$$Q_j(\lambda) = \mu_j(\lambda)W_j \left[\frac{r_j - 1}{r_j} \right] \omega_j g_j$$

$$Q_i(\lambda_j) = \mu_i(\lambda_j)W_i \left[\frac{r_i - 1}{r_i} \right] \omega_i g_i$$

The data listed in Table 7-2 are taken from a study by Shiraiwa and Fujino,[4] and these illustrate the relative magnitudes of primary, secondary and tertiary fluorescence. The Ni/Fe/Cr system is one of the more complex element combinations encountered and in this system all three types of fluorescence generally have to be considered. Data have been selected in three groups of three, in the first group the iron concentration is constant, in the last group the nickel concentration is constant and in the middle group the concentration of all three elements vary. Enhancement by iron (the best "enhancer" of chromium) generally contributes about 15–25% of the total chromium count rate. The *total* effect of nickel (i.e., direct enhancement plus third element) is always less than 10%, of which the tertiary (third element) fluorescence is never more than a few percent.

TABLE 7-2.
Measurement of Cr Kα intensity in Ni⁺ Fe⁺ Cr alloys†

Composition %			Total Cr Kα intensity (count ratio)	Relative Cr Kα Primary excitation	Expressed as percentage of total		
					Enhancement		Third
Ni	Fe	Cr			by Fe	by Ni	element (Ni)
15	75	10	0.145	71.9	24.8	2.1	1.2
10	75	15	0.211	73.9	24.2	1.2	0.7
5	75	20	0.273	75.7	23.5	0.6	0.4
40	50	10	0.135	74.3	17.1	6.2	2.4
20	65	15	0.205	74.9	21.0	2.8	1.3
10	70	20	0.269	76.3	21.6	1.5	0.6
25	65	10	0.141	72.9	22.0	3.3	1.8
25	60	15	0.202	75.4	19.3	3.7	1.6
25	55	20	0.259	77.5	17.0	4.0	1.5

† Data taken from Shiraiwa, T. and Fujino, N., *Bull. Chem. Soc. Japan*, **40**, 2289 (1967).

Although the agreement between experimental and theoretical results are good in this instance, three major factors have inhibited the use of these basic algorithms for the direct calculation of concentration. The first of these is that the complexity of the algorithm requires the availability of large computing facilities, probably something like forty thousand words of direct access core would be required. The second problem is that currently available fundamental constants are barely good enough to give the accuracies required, although this situation will undoubtedly change over the next few years.‡ The last difficulty is the problem which arises in the calculation of the output from the X-ray tube—the $J(\lambda)$ integral in equation (3-9).

‡ The most recent review at time of going to press was Bambynek *et al.*, *Rev. Mod. Phys.* **44**, 716 (1972).

Although expressions are available for the prediction of X-ray output from an infinitely thick target, considerable difficulties arise in attempting to correct the spectrum for self absorption, Compton shift, and absorption by the X-ray tube window. A further complication arises in that although the X-ray tube target can be considered infinitely thick for the impinging electrons, it is certainly not infinitely thick for the white X-radiation that is produced. For example, a typical Cr anode is of the order of $50 \pm 5 \mu m$ thick and this thickness variation is sufficient to modify the tube output. In practice it is better to actually measure the tube output than to attempt to calculate it.

The following sections show the various modifications to the basic algorithm, given in Table 7-1, which have been employed in actual routine analysis. It will be seen that much simpler expressions can be derived by introducing certain assumptions, but although this makes the data handling simpler, it also requires more calibration standards, and tends to reduce the concentration range over which the modified algorithm is applicable.

7.3 THE FUNDAMENTAL PARAMETERS APPROACH

Probably the most successful approach to the use of fundamental data for the evaluation of concentrations from measured intensities is the "fundamental parameters" method due to Criss and Birks.[5] In their method the basic algorithm is similar to that given in Table 7-1, again the tertiary fluorescence term being ignored. In order to overcome the difficulty of calculating the output of the X-ray tube from basic theory, the tube spectrum is measured by a separate experiment.[7] Figure 7-1 shows the configuration of the experimental set-up for this measurement and illustrates a typical spectrum obtained from a chromium anode tube operated at 45 kV constant potential.

In order to allow simple mathematical expression of the integral form of the tube spectrum, the measured spectrum is tabulated in the form of intensity $I(\lambda)$ per 0.02 Å interval, $\Delta\lambda$, starting at a short wavelength limit equivalent to λ_{min} (i.e. the short wavelength limit of the tube). Figure 7-2 shows the form of the summation. In order to allow easily interchangeable values of λ_{edge} (i.e., the absorption edge value of the measured element), an extra term $D_i(\lambda_i)$ is added to the summation term. This is set to 0 for values of $\Delta\lambda > \lambda_{edge}$ and to unity for values $< \lambda_{edge}$. The integral form of $J(\lambda)$ is then replaced by $\sum D_i(\lambda_i)I(\lambda)\Delta\lambda$. For example, the primary fluorescence term in Table 7-1 would be of the form

$$I(\lambda_i) \text{ primary} = W_i K_i \sum_i \frac{D_i(\lambda_i)I(\lambda)\Delta\lambda\,\mu_i(\lambda)}{W_j\alpha_j} \qquad (7-1)$$

where

$$K_i = \frac{w_i g_i[(r_i - 1)/r_i]}{\sin\psi_2} \qquad (7-2)$$

and α_j is the same as defined in equation (3-8).

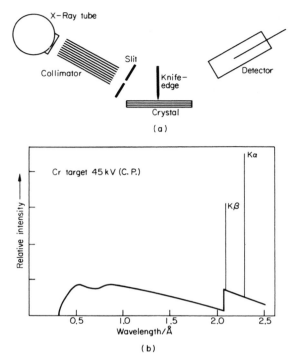

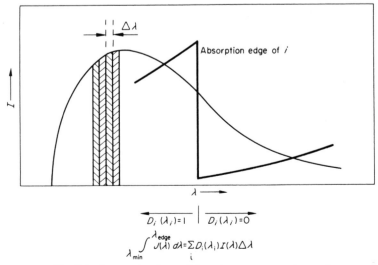

Fig. 7-1   Experimental arrangement (a) for the measurement of X-ray tube spectra and the output
(b) for a chromium anode tube

$$\int_{\lambda_{min}}^{\lambda_{edge}} J(\lambda)\, d\lambda = \sum_i D_i(\lambda_i) I(\lambda) \Delta\lambda$$

Fig. 7-2   Summation form of the primary spectrum integral

There is a similar form which also includes secondary fluorescence. By utilizing intensity ratio R_i, relative to a reference standard, the K_i term is eliminated, leaving only terms in R and W, plus many constants. Use of ratios will also correct for certain geometric factors, for example, the incident angle ψ_1 is in practice, a range of angles, due to the divergence of the primary X-ray beam.

It will be noted from equation (7-1) that X-ray intensity is expressed in terms of total element composition. Since it is the elemental composition that is being sought in the first place, it is obvious that some form of iterative procedure[8] is necessary to arrive at a complete solution. In the Criss–Birks procedure, linear relationships between normalized R_i's and W_i's are assumed and from these (incorrect) W_i's, a new set of R_i's is calculated. The differences between measured and calculated R_i's are then used to correct the W_i's and the procedure started all over again. Three or four such iterations are normally sufficient. Table 7-3 shows data reported by Criss and Birks on Hastelloy X and it will be seen that even though seven iterations were carried out, four would have given sufficient precision.

TABLE 7-3.
Composition of Hastelloy X calculated by the "fundamental parameters", approach†

Iteration	%Cr	%Fe	%Co	%Ni	%Mo
1	26.81	18.77	1.41	39.68	13.33
2	22.90	19.60	1.47	46.07	9.96
3	22.11	18.59	1.44	47.55	10.31
4	22.20	18.38	1.43	47.58	10.41
5	22.23	18.39	1.43	47.53	10.42
6	22.23	18.40	1.43	47.53	10.41
7	22.23	18.40	1.43	47.53	10.41

† Data from reference 5.

The great advantage of the fundamental parameters approach is its wide applicability and, in principle, its freedom from the requirement for large numbers of standards. As previously stated, its major disadvantage lies in the need for substantial computing facilities.

7.4 THE CONCEPT OF EFFECTIVE WAVELENGTH

From what has been discussed thus far, it will be apparent that the major stumbling blocks in the calculation of elemental composition from characteristic line intensities are the expression of the X-ray tube distribution and the estimation of its effectiveness in exciting the various characteristic wavelengths. One practical means of circumventing this difficulty is to consider a hypothetical "effective" wavelength which, as far as excitation is concerned, would have the same "effectiveness" as the X-ray tube continuum.[1] This concept is perhaps not too hard to accept when one recalls from Section 3.2 (p. 40) that the most effective portion of the continuum is that immediately to the short wavelength side of the absorption edge of the excited element, and further, the excitation efficiency of a given portion of the spectrum decreases as it moves further away from the edge. In general terms, for a given element measured in a fixed equipment configuration, the efficiency of a given wavelength λ (in the tube spectrum) in exciting a wavelength λ_i is given by

$$C(\lambda\lambda_i) = \frac{\mu_i(\lambda)}{\sum_j W_j \alpha_j} \qquad (7\text{-}3)$$

The factor $C(\lambda\lambda_i)$ is the so-called efficiency factor introduced by Spielberg.[9] Figure 7-3 gives a graphical interpretation of the effective wavelength concept. (a) and (b) show the X-ray tube distribution and absorption function of element i.

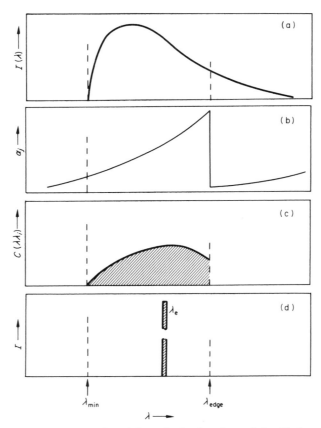

Fig. 7-3   Graphical interpretation of the excitation function and the effective wavelength

(c) shows the curve relating the efficiency factor with wavelength, the so-called probability function. (d) shows the effective wavelength λ_e which is essentially the weighted mean of the $C(\lambda\lambda_i)$ curve. In practice, it is found that where characteristic tube lines are unimportant in the excitation process, λ_e is approximately equal to 2/3 of the absorption edge value of the excited element. This will generally be the case where only L or M characteristic tube lines occur to the short wavelength side of the absorption edge of the excited element. Where K lines from the X-ray tube target element occur to the short wavelength side of the edge, λ_e is taken as being equal to the Kα wavelength of the X-ray tube target element.

It can happen that the curve of α_j against λ is not a smooth curve[10] and where this is the case λ_e no longer has a constant value and may vary by several percent. For example, Fig. 7-4 illustrates an extension of Sherman's work, published by Alvarez and de Vries,[11] on the $Fe_2O_3 + ZnO$ system. Here the tube spectrum and the absorption curve for iron (dotted line) are shown, along with the probability functions for 100%, 90% and 30% Fe_2O_3 in ZnO. The modification in the shape of the probability function due to the zinc absorption edge at 1.28 Å is clearly seen, and this change in the shape of the distribution causes a shift in the effective wavelength from 1.19 Å to 1.28 Å over the range 100 to 10% Fe_2O_3. These values should be compared with the constant value of 1.20 Å proposed by Sherman. As previously mentioned, this effect may be even more marked in cases where the characteristic lines from the X-ray tube play an important part in the excitation.[12]

The subject of the accuracy of the effective wavelength concept still invokes considerable discussion,[13] but nevertheless it has been found extremely useful in practice, as will be shown in the following section.

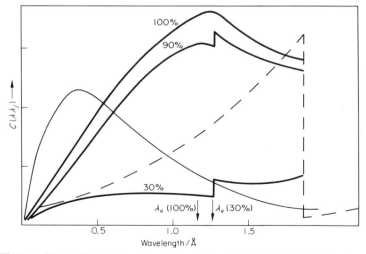

Fig. 7-4   Probability functions for the excitation of FeKα in Fe_2O_3/ZnO mixtures

7.5   ALGORITHMS BASED ON THE USE OF EFFECTIVE WAVELENGTH

An algorithm which utilizes the effective wavelength concept will be of a similar form to that given in Table 7-1, the difference being the replacement of the integral form of $J(\lambda)$ by the effective wavelength λ_e. From this point on, the handling of the calculation is similar to that of the fundamental parameter method. One such program which has made use of this approach is the FORTRAN program CORSET, developed by Stephenson.[14] The CORSET program makes use of both primary and secondary fluorescence corrections, but ignores the tertiary fluorescence term. The program has been found applicable to a wide range of matrices including rocks, glasses and high alloyed steels and the obvious advantage of this procedure lies in its ability to apply necessary corrections without any knowledge of the spectral distribution of the X-ray source.

7.6   FURTHER SIMPLIFICATIONS OF THE INTENSITY/CONCENTRATION ALGORITHM

A very common use of the effective wavelength concept—although this may not seem obvious at first sight—is to be found in the much used expression

$$I(\lambda_i) = \frac{K_i W_i}{\sum_j W_j \alpha_j} \tag{7-4}$$

i.e., the concentration of a given element i in a matrix is directly proportional to the intensity of one of its characteristic wavelengths, which is in turn inversely proportional to the total matrix absorption.

The origin of the expression given in equation (7-4) will be seen to be the primary fluorescence term in Table 7-1 in which the following substitutions have been made:

(a) considering only primary fluorescence

$$I_p(\lambda_i) = I(\lambda_i)$$

(b) assuming an effective wavelength λ_e

$$\lambda_e = \int_{\lambda_{min}}^{\lambda_{edge}} J(\lambda)\, d\lambda$$

(c) For a given element i measured under fixed conditions of excitation, geometry and dispersion

$$\frac{Q_i(\lambda)}{\sin \psi_2} = \text{constant} = K_i$$

(d) The total sample absorption is given by

$$\mu_s(\lambda) \sin \psi_1 + \mu_s(\lambda_i) \sin \psi_2 = \sum_j W_j[\mu_j(\lambda) + A\mu_j(\lambda_i)]$$

or, as previously defined,

$$\sum_j (W_j \alpha_j)$$

Substitution of these terms in the primary fluorescence expression gives the form in equation (7-4).

The usual procedure when using this expression is to determine K_i by calibration or to use the ratio form

$$\frac{C_i^x}{C_i^s} = \frac{I^x(\lambda_i)}{I^s(\lambda_i)} \cdot \frac{(\sum_j W_j \alpha_j)^s}{(\sum_j W_j \alpha_j)^x} \tag{7-5}$$

where x and s refer respectively to unknown specimen and reference sample.

The slope of a calibration line, i.e., $1/K_i$, will be defined by the ordinates $C_i^x = 0$, $I^x(\lambda_i) = 0$, and $C_i^x = C_i^s$, $I^x(\lambda_i) = I^s(\lambda_i)$, see Fig. 7-5. The slope of the curve will also be fixed by the absorption of s for λ_i and the condition for a linear relationship between C_i^x and $I^x(\lambda_i)$ will obviously be

$$\left[\sum_j W_j \alpha_j \right]^s = \left[\sum_j W_j \alpha_j \right]^x$$

i.e., that there is no difference in the total absorption of samples over the range of the calibration. Although in practice this idealized situation is never completely realized, it is often found to be approximately true. In other words, provided that the total absorption of the specimens do not change significantly

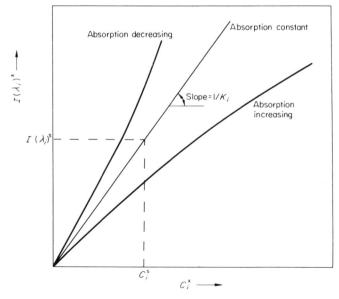

Fig. 7-5   Shapes of calibration curves due to absorption phenomena

over a given (and generally limited) concentration range a linear relationship between measured intensity and concentration can be assumed. This is the basis of so-called "In-type" analysis where calibration standards are used to ascertain the slope of the curve as well as to ensure its constancy.

7.7 SECONDARY ABSORPTION CORRECTIONS

As previously mentioned, the secondary absorption term is generally the dominant factor in the total absorption of the specimen. Where this is the case, a further simplification can be made to equation (7-4) i.e.,

$$C_i = \frac{I(\lambda_i)K_i}{\sum_j W_j\mu_j(\lambda_i)} \tag{7-6}$$

where the term $\sum_j W_j\mu_j(\lambda_i)$ is simply the total secondary absorption of the matrix for the wavelength λ_i. Where this total absorption can be measured by a separate experiment, an absorption correction can be readily applied.

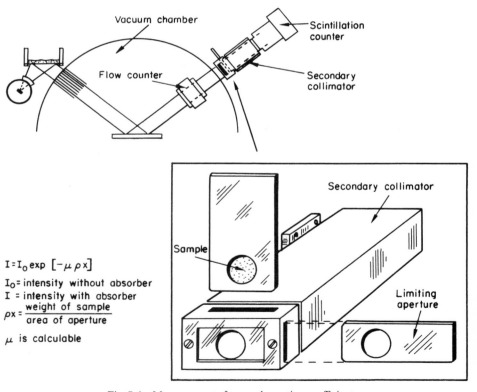

$I = I_0 \exp\left[-\mu\rho x\right]$

I_0 = intensity without absorber

I = intensity with absorber

$\rho x = \dfrac{\text{weight of sample}}{\text{area of aperture}}$

μ is calculable

Fig. 7-6   Measurement of mass absorption coefficients

Figure 7-6 illustrates the basis of such an experiment to measure the total secondary absorption of a specimen.[15-17] A known weight (usually around 0.6 g) of sample is pressed into a plastic holder of known cross-sectional area. Since both total weight and cross-sectional area are known, it is possible to calculate the mass-thickness ρ_x of the sample thus

$$\rho x = \frac{\text{weight of sample}}{\text{area of hole}}$$

By measuring the intensity of the measured wavelength from a suitable sample with (I) and without (I_0) the absorber in position, it is possible to calculate the mass absorption coefficient μ by use of the Beer–Lambert equation

$$I = I_0 \exp(-\mu\rho x)$$

Table 7-4 gives typical results for a series of synthetic samples containing small amounts of strontium, and the calculated and experimentally determined μ values for strontium Kα are given.

TABLE 7-4.
Measurement of secondary absorption for Sr Kα by means of the absorption cell†

S/N	%SiO$_2$	%CaCO$_3$	μ calculated	μ experimental
1	50.0	50.0	10.1	9.9
2	83.3	16.7	7.8	7.8
3	100.0	0	7.2	6.1
4	16.7	66.7	12.5	13.0
5	33.3	83.3	11.4	11.2
6	0	100.0	13.6	12.7
7	66.7	33.3	9.2	8.7

† All samples contain traces (5–250 ppm) of strontium.

Absorption corrections are applied using the form

$$C_x = C_s \frac{I_s}{I_x} \frac{\mu_s^T}{\mu_x^T}$$

where C_s is the concentration of the measured element in a standard reference (or a calibration curve) which has a total absorption equal to μ_s^T and which gives an intensity I_s. C_x is the concentration of the same element in the unknown specimen, μ_x^T its measured total absorption and I_x the measured intensity.

This absorption correction method finds great use where reasonably short wavelengths (<1.5 Å) are to be measured. It is not so useful for longer wavelengths, where absorption coefficients are generally relatively larger because of the practical difficulties involved in obtaining a specimen thin enough to transmit a measurable number of X-ray photons.

Use can sometimes be made of the fact that the scattering power of a specimen is related to its total absorption. Anderman[18] has proposed the use of a scattered characteristic line from the X-ray tube spectrum as an internal standard.

7.8 EMPIRICAL CORRECTION PROCEDURES

Great use has been made in quantitative X-ray spectrometry of so-called "empirical" correction procedures. In practice these empirical procedures are of several types, some of which really are not empirical at all, but which are based on the fundamental equations previously discussed. Two major classes exist, the first of these being truly empirical algorithms in which the concentration–intensity relationship is expressed as a polynomial and then multiple regression techniques employed to solve for the constant terms, using well analysed standards as a source of raw data. The second major class involves a semi-empirical approach in which the concentration–intensity algorithm is expressed in such a form that the constant and correction terms have significance in terms of actual absorption coefficients, and indeed these constant terms are calculable from mass absorption coefficient data. In practice it is generally the case that, as with the truly empirical relationships, multiple regression techniques based on well analysed standards are employed to determine the constant terms. But the distinction between the classes remains, in that in the second class the correction constants *are* calculable.

In the interest of continuing a logical development of modifications to the basic intensity–concentration algorithm the second class, the semi-empirical correction methods will be discussed first.

7.9 SEMI-EMPIRICAL METHODS

From equation (7-4) it will be seen that for a binary mixture A, B, the following relationship holds.

$$I_A = \frac{K'_A W_A}{W_A \alpha_A + W_B \alpha_B} \tag{i}$$

where I_A is the intensity of a selected wavelength of element A and W_A and W_B are the respective weight fractions of A and B. α_A and α_B are the total absorption due to elements A and B. K'_A is a constant for element A measured under fixed conditions. Since $W_A + W_B = 1$ then $W_A = 1 - W_B$

substitute in (i)

$$I_A = \frac{K'_A W_A}{(1 - W_B)\alpha_A + W_B \alpha_B} = \frac{K'_A W_A}{\alpha_A + W_B(\alpha_B - \alpha_A)} \tag{ii}$$

let

$$\alpha_{AB} = \frac{\alpha_B - \alpha_A}{\alpha_A} \quad \text{therefore} \quad \alpha_B - \alpha_A = \alpha_{AB}\alpha_A$$

substitute in (ii)

$$I_A = \frac{K'_A W_A}{\alpha_A + W_B \alpha_A \alpha_{AB}} \tag{iii}$$

let $K'_A = \alpha_A/K_A$

substitute in (iii)

$$I_A = \frac{W_A/K_A}{1 + \alpha_{AB} W_B}$$

or

$$W_A = I_A K_A [1 + \alpha_{AB} W_B] \tag{iv}$$

Since for a pure sample of element A, $W_A = 1$ and $W_B = 0$, then

$$K_A = 1/I_A^P$$

where I_A^P is the intensity of the measured wavelength from the pure element. If R_A is taken as the ratio of the sample intensity to the pure element intensity, i.e., I_A/I_A^P, substitution in (iv) gives

$$W_A = R_A[1 + \alpha_{AB} W_B] \tag{7-7}$$

Note that

$$\alpha_{AB} = \frac{\alpha_B}{\alpha_A} - 1 = \left[\frac{\mu_B(\lambda) + A\mu_B(\lambda_A)}{\mu_A(\lambda) + A\mu_A(\lambda_A)}\right] - 1 \tag{v}$$

also

$$K'_A = \frac{\alpha_A}{K_A} = \frac{\alpha_A \sin \psi_2}{Q_A(\lambda)} = \frac{[\mu_A(\lambda) + A\mu_A(\lambda_A)] \sin \psi_2}{\omega_A g_A[(r_A - 1)/r_A]} \tag{vi}$$

The constant α_{AB} is a factor (the so-called alpha factor) which represents the effect of element A on element B and there are three ways by which it may be determined. The following illustrates these three methods, by reference to a series of data taken on Fe + Cr binary alloys in which iron is taken as the measured element. The measured intensity and concentration data for these binaries are given in Table 7-5.

(a) Graphical method

Equation (7-7) is a linear expression and can be rewritten in the form of a straight line, i.e., $y = mx + b$, as follows.

$$W_A/R_A = \alpha_{AB} W_B + 1$$

Thus by plotting W_A/R_A against W_B, a straight line relationship should be obtained of slope α_{AB} with an intercept of unity. Figure 7-7 shows such a plot obtained with the data given in Table 7-5. The graphical method is additionally useful in that it allows an immediate visual check of the validity of equation (7-7) over the concentration range in question.

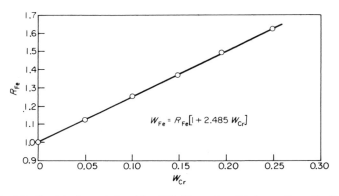

Fig. 7-7   Straight line relationship of W_A/R_A against W_B for an Fe + Cr alloy

TABLE 7-5.
Composition and intensity data on Fe + Cr binary alloys

Sample	W_{Fe}	W_{Cr}	I (Fe Kα) /(10^4 counts/s)	I (Cr Kα) /(10^4 counts/s)	R (FeKα)
1	1.0000	0	1.8564	0	1.0000
2	0.9506	0.0494	1.5703	0.3031	0.8459
3	0.9001	0.0999	1.3358	0.5655	0.7194
4	0.8518	0.1482	1.1553	0.7916	0.6223
5	0.8050	0.1950	1.0034	0.9993	0.5404
6	0.7509	0.2491	0.8608	1.2091	0.4635

(b)   Regression analysis

Since the essence of the problem is to obtain the best fit of experimental data to a given algorithm, one obvious approach is to use a standard regression procedure to obtain the required fit. Most computer houses supply regression programs within their software packages and Fig. 7-8 gives a typical output obtained when the data from Table 7-5 were fitted to equation (7-7). In order to ensure that the calibration line goes approximately through the point $R = 0$, $W = 0$ (a fact which *we* know but the computer does not) an extra data set has been added (S/N 7). The actual value of the background is given under the $B(I)$ column. The output from the regression analysis includes the (entered) chemical

```
ELEMENT  01   FE
         CHEMICAL    CALCULATED    DIFFERENCE

01       1.0000      1.0004        -.0004
02        .9506       .9505         .0001
03        .9001       .8992         .0009
04        .8518       .8527        -.0009
05        .8050       .8037         .0013
06        .7509       .7520        -.0011
07        0           *             .0000

   SIGMA             .00110
```

$$k(I) \qquad\qquad B(I)$$
$$.53892 \qquad\qquad -.00005$$
$$\alpha(I,J) \qquad\qquad B(I,J)$$
$$1.34393$$

Fig. 7-8

analyses, the calculated X-ray values and the differences. Also given under "SIGMA" is the standard deviation of these differences. Finally, the values are given for the constants $k(I)$, $\alpha(IJ)$ and $B(I)$. Thus the equation of the line in this case is

$$W_{Fe} = I_{Fe}(0.53892 + 1.34393 W_{Cr}) - 0.00005$$

note

$$\alpha_{Fe/Cr} = \frac{1.34393}{0.53892} = 2.494.$$

(c)   Calculation from absorption coefficients

As was shown in the expression given in equation (v) the alpha factor can be calculated from absorption coefficient data. In this instance

$$\alpha_{Fe/Cr} = \left[\frac{\mu_{Cr}(\lambda) + A\mu_{Cr}(Fe\ K\alpha)}{\mu_{Fe}(\lambda) + A\mu_{Fe}(Fe\ K\alpha)}\right] - 1$$

The wavelength λ is the effective wavelength for Fe Kα which, as was stated previously, can be taken as 2/3 of the absorption edge value of the excited element—in this instance Fe Kα. Since the absorption edge value of iron is 1.744 Å, λ in this instance is equal to about 1.16 Å. The following data were taken from absorption tables.[19] using the value 1.937 Å for the Fe Kα line.

$$\mu_B(\lambda) = \mu_{Cr}(1.16 \text{ Å}) = 112$$

$$\mu_B(\lambda_A) = \mu_{Cr}(1.927 \text{ Å}) = 471$$

$$\mu_A(\lambda) = \mu_{Fe}(1.16 \text{ Å}) = 140$$

$$\mu_A(\lambda_A) = \mu_{Fe}(1.937 \text{ Å}) = 72$$

The value of the geometric factor A was in this instance 1.65 thus

$$\alpha_{Fe/Cr} = \left[\frac{112 + (1.65 \times 471)}{140 + (1.65 \times 72)} \right] - 1 = 2.44$$

The agreement between the three values of $\alpha_{Fe/Cr}$ ix about 2%, which is as good as can be expected in view of experimental errors and uncertainties in the absorption coefficient values. In practice, this is generally sufficient accuracy since it is the effect of this correction term relative to unity which determines the accuracy of the concentration value. Since W_B (the weight fraction of the interfering element) is generally at the very most 0.5, the size of the correction term is reduced accordingly. The α coefficients can also be calculated from the full basic intensity equation given in Table 7-1, and this is the scheme of the ALPHAS program proposed by de Jongh.[25] The major practical advantage to be offered by this approach is the ability to be able to *calculate* the α coefficients using a large off-line computer, but to *employ* them in a relatively simple algorithm, such as that given in equation (7-9), using a small on-line computer.

7.10 APPLICATION OF SEMI-EMPIRICAL CORRECTIONS IN MULTI-ELEMENT ANALYSES

The treatment given for a binary mixture can be similarly employed for multicomponent mixtures since in the latter case, A still represents the matrix element but B now represents the rest of the matrix. Thus the α_{AB} term does not involve just W_B but all interfering elements (very rarely would this imply *all* matrix elements). Thus in the expression given in equation (7-7) the $\alpha_{AB}W_B$ correction term is replaced by a summation term involving all intefering elements j. Thus

$$W_i = R_i[1 + \sum_j \alpha_{ij}W_j] \qquad (7\text{-}8)$$

R_i it will be remembered, is a ratio term which refers to the pure element i. In practice, it is often simpler to split the R_i term as it was in equation (iv) and to carry the K_i term inside the bracket giving the general form

$$W_i = R_i'[k_i + \sum_j \alpha_{ij}'W_j] + B_i \qquad (7\text{-}9)$$

where R_i' now represents an intensity ratio term against a reference standard which need not necessarily be a pure specimen of element i. B_i is a background term (having the dimensions of concentration). Equation (7-9) is the generalized form of the Lachance–Traill equation[20] which is the one used previously in the regression analysis to find $\alpha_{Fe/Cr}$. Note that a simple relationship exists between the terms in equations (7-8) and (7-9) namely

$$R'_i k_i = R_i \quad \text{and} \quad \alpha_{ij} = \alpha_{ij}/k_i$$

So far only the effects of primary and secondary absorption have been considered in the development of the alpha correction model. Where enhancement effects are present the determined alphas will occur as negative values. Where regression procedures are employed to determine the constants it makes little difference whether absorption or enhancement effects are present. However, the *calculation* of the α_{ij} terms becomes more difficult where enhancement effects are significant and although the calculation is not too difficult, it is not the type of calculation that one would normally carry out by hand. A good rule of thumb for approximate calculations, is that where a strongly enhancing element is present (an element with a wavelength *very* close to the absorption edge of the analysed element) a constant value of about 0.5 should be subtracted from the α_{ij} term, which has been calculated from absorption data, as shown previously. The value of 0.5 should be reduced to around 0.3 where elements of lower enhancing ability are present.

It will be noted that the alpha correction procedure so far described will yield a series of simultaneous equations in multi-element systems since concentration terms occur on both sides of the expressions.

A further simplification can be used to reduce the form of equation (7-9) to a linear expression involving a summation of interfering element *intensities* and correction factors. The basis is the assumption that a linear relationship exists between the concentration and intensity of the interfering elements j. Hence, from equation (7-4),

$$I_j = K_j W_j \tag{7-10}$$

substitution in equation (7-9) gives the form

$$W_i = R'_i[k_i + \sum_j k_{ij} I_j] + B_i \tag{7-11}$$

k_{ij} is a constant term representing the effect of element j on element i, but it should be differentiated from the α_{ij} term. The k_{ij} notation is used where the *intensity* of the interfering element j is used to correct for its effect on the analysed element i. The α_{ij} notation is used where the *concentration* of element j is used for correction. Equation (7-11) is a general expression of the Lucas–Tooth/Price formalization.[21] This algorithm offers the advantage of simplicity of use because of the ease of handling of the right hand side of the equation. It has the disadvantage that it is applicable only over relatively limited concentration (or better limited absorption) ranges. Errors in W_i will start to accrue whenever the linear relationship between I_j and W_j breaks down. As in the case of the concentration correction model, provided that the total correction term $\sum_j k_{ij}I_j$ is small relative to k_i (generally less than 25 % of the value of k_i) the error in the correction term can be significantly greater than the required accuracy in W_i.

The data listed in Table 7-5 can be used to illustrate the comparison between concentration and intensity type corrections. Rewriting equation (7-11) in the form used in equation (7-7) but using an intensity term, one obtains:

$$W_{Fe} = R_{Fe}[1 + k_{Fe/Cr}I_{Cr}] \tag{7-12}$$

As was done before for the concentration correction model in Fig. 7-7, equation (7-12) can be rearranged to give an equation in the form of $y = mx + b$. Figure 7-9 shows the plot of $(W/R)_{Fe}$ against I_{Cr} which again should give a straight line of slope k_{ij} and an intercept of unity. In this instance, however, it is impossible to obtain a linear fit of the data. A *reasonably* linear fit can be obtained by reducing the concentration range, and the indicated line matches the points quite well over the 70–90 % range of iron concentration.

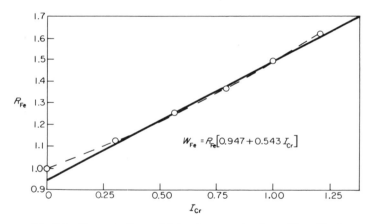

Fig. 7-9   Concentration and intensity correction for an Fe + Cr alloy

From the foregoing it will be apparent that the intensity type correction (ΔI) is applicable over much smaller concentration ranges than the concentration type correction (ΔC). Although the latter will, in general, not be as universally applicable as the fundamental approach, it is possible to apply the method with a relatively small computing capability (typically $8K$ of direct access core, with a 16 bit word length).

7.11   EMPIRICAL POLYNOMIALS

It is difficult to draw the line between the semi-empirical relationships already discussed and purely empirical polynomials. As has been previously stated, the major criterion for separating the two categories is the ability or non-ability to calculate correction factors from first principles. The ΔC correction method is obviously a case where the α_{ij} correction factors are calculable, but on closer inspection it will be seen that the ΔI type correction method also, in principle, allows the calculation of the k_{ij} factors. In the latter case, the k_{ij} is related to the α_{ij} by the factor K_j (the reciprocal of counts/s per $\%$ for element j). However, one very important condition in the application of either of these methods, or indeed for any of the fundamental methods, is that the measured intensities are free from instrumental artifacts such as dead time, line overlap, etc. This introduces another criterion into the categorization of the empirical methods and it is generally found that, when a higher order polynomial expression is employed, no attempt is made to separate out effects such as dead time. Typical of a high order polynomial is that suggested by Alley and Myers[22] which expresses the weight fraction W_i of the ith component with its observed X-ray intensity R_i relative to the pure element:

$$W_i = \sum_{i=1}^{n} \beta_i I_i + \sum_{\substack{i=1 \\ i<j\leqslant n}}^{n-1} \beta_{ij} I_i I_j + \varepsilon_i \qquad (7\text{-}13)$$

in which β_i, β_{ij} and ε_i are regression coefficients to be estimated by ordinary least squares techniques. This technique and others like it are based on Scheffe's "canonical" polynomial for second degree equations.[23] Although this approach has been used with some success for the analysis of multi-component mixtures,[22,24] it is doubtful whether it will prove more useful than the semi-empirical methods which, if nothing else, allow a relatively simple "ball-park" check to be made of the inter-element coefficients.

References

(1) Sherman, J., *Spectrochim. Acta* 7, 283 (1955).
(2) Sherman, J., *Spectrochim. Acta* 11, 466 (1959).
(3) Shiraiwa, T. and Fujino, N., *Japan. J. Appl. Phys.* 5, 886 (1966).
(4) Shiraiwa, T. and Fujino, N., *Bull. Chem. Soc. Japan* 40, 2289 (1967).
(5) Criss, J. W. and Birks, L. S., *Anal. Chem.* 40, 1080 (1968).
(6) Gould, R. W. and Bates, S. R., *X-Ray Spectrom.* 1, 29 (1972).
(7) Gilfrich, J. V. and Birks, L. S., *Anal. Chem.* 40, 1077 (1968).
(8) Criss, J. W. and Birks, L. S., in McKinley, T. D., Heinrich, K. F. J. and Wittry, D. B. (Eds.), *The Electron Microprobe*, Wiley, New York, 1966, p. 217.
(9) Spielberg, N., *Philips Res. Rep.* 14, 215 (1959).
(10) Jenkins, R. and de Vries, J. L., *Worked Examples in X-ray Analysis*, 2nd Edn., problem 34, Macmillan, London, 1970.
(11) Alvarez, A. G. and de Vries, J. L., *Proceedings of the Conference on X-ray Analysis*, Swansea, Philips, Eindhoven, 1966, p. 16.
(12) Müller, R., *Spectrochim. Acta* 18, 123, 1515 (1962).
(13) Stephenson, D. A. and Tertian, R., *Spectrochim. Acta* B27, 153 (1972).
(14) Stephenson, D. A., *Anal. Chem.* 43, 1761 (1971).
(15) Salmon, M. L. and Blackledge, J. P., *Norelco Rept.* 3, 68 (1956).
(16) Salmon, M. L., *Advan. X-ray Anal.* 2, 305 (1958).

(17) Norrish, K. and Taylor, R. M., *Clay Minerals Bull.* **5**, 98 (1962).
(18) Anderman, G. and Kemp, J. W., *Anal. Chem.* **30**, 1306 (1958).
(19) Reynolds, R. C., Mass Absorption Tables, in *Handbook of X-ray and Microprobe Data*, Dewey, Mapes and Reynolds, Polycrystal Book Service, Pittsburgh, 1967, pp. 323–339.
(20) Lachance, G. R., and Traill, R. J., *Can. Spectry.* **11**, 43 (1966).
(21) Lucas-Tooth, H. J. and Price, B. J., *Metallurgia* **54**, 149 (1961).
(22) Alley, B. J. and Myers, R. H., *Norelco Rept.* **15**, 87 (1968).
(23) Scheffe, H. and Roy, J., *Statistical Soc. Series B*, **20**, 344 (1958).
(24) Stephenson, D. A., *Anal. Chem.* **43**, 310 (1971).
(25) de Jong, W. K., *X-Ray Spectrom.* **2**, 151 (1973).

CHAPTER 8

the study of chemical bonding

8.1 INTRODUCTION TO THE USE OF X-RAY SPECTROSCOPIC METHODS FOR THE STUDY OF CHEMICAL BONDING

One of the major areas of research over the past decade has been that involving the use of X-ray spectroscopic methods for the study of chemical bonding. In the early 1960's, the major areas of interest were those of wavelength shift and band spectra of (mainly) the lower atomic number elements, plus the absorption edge fine structure of (mainly) the transition series elements. After the rapid growth of electron spectroscopic methods in the late 1960's and early 1970's these techniques have now largely superseded the older ones. Nevertheless, inasmuch as the types of data obtained are not identical, each of the methods about to be described has its own areas of application. Although the "older" methods involving X-ray emission and absorption edge fine structure could reasonably be called "X-ray" methods, the "newer" techniques, included under the general title of ESCA (electron spectroscopy for chemical analysis), are PES (photoelectron spectrometry), Auger spectroscopy and ultraviolet excited electron spectroscopy, are not X-ray methods, although an X-ray source may be involved. Although the electron spectroscopic methods might apparently fall outside the scope of this book, some mention of them must be made, specifically in the context of comparing and contrasting them with X-ray methods, since this will offer a further insight into our knowledge of electronic levels and binding energies.

The following description of electron spectroscopic methods is relatively cursory and for detailed information regarding these techniques, the reader is referred to other works dealing specifically with this field of application.[1-4] The books by Siegbahn[1] and Sevier[3] are particularly useful.

Table 8-1 briefly summarizes the features, advantages and disadvantages of the five basic methods for the study of chemical bonding. It will be seen that the major differences are those of effective sample penetration and the binding levels which are excited. Thus X-ray emission gives only information on bonding in outer orbitals whereas PES can be made to excite *any* orbital. Ultraviolet source electron spectrometry is only applicable to outer levels since it is capable

TABLE 8-1.
Use of X-ray methods for the study of chemical bonding

Method	Features	Disadvantages	Uses
X-Ray emission	Simple interpretation. Dependent upon electron density in outer orbitals. Penetration 1–10 μm	Only useful for solids. Very limited range of application ($Z = 11$–17)	Chemical bonding studies, structural chemistry
PES	Sharp lines, good resolution. Solids, liquids or gases. Measures bonding energies in selected orbitals. Penetration 10 Å–1 μm	Difference between applied and surface potential demands use of careful calibration. Ultimate resolution depends on natural line width of source (0.7–4 eV)	Organic and inorganic structure analysis, biochemistry, catalysis studies, corrosion studies, etc.
Auger	Very thin layer contributes, 5–100 Å	Complex spectra. High background due to scatter of primary electrons. Very high vacuum required	Surface phenomena, diffusion studies, thin film studies
Ultraviolet source photoelectron spectrometry	Outer orbitals excited. Ultra high resolution	Only gases or low vapour pressure liquids	Study of bonding in gaseous phase
Absorption edge fine structure	Relatively large volume of sample irradiated. Penetration depth 10–1000 μm	Interpretation rather empirical. Specimen preparation difficult. Intensity depends on μ_x rather than μ	Catalysis studies, general inorganic applications

of excitation energies *only* of the order of a few tens of electron volts. Auger electrons are always produced during irradiation but since the ratio of Auger electrons to X-ray photons increases rapidly with decrease of atomic number, the Auger method is more applicable to the lower atomic number elements. Both Auger and absorption edge spectroscopy are difficult to interpret and the data obtained do not lend themselves readily to confirmation by calculations from first principles.

It will become apparent that there is much overlap in the areas of application of these various techniques and a better classification is really required. The terms PESIS (photoelectron spectrometry for *inner* shells) and PESOS (photoelectron spectrometry for *outer* shells) have been suggested[7] and although at the time of writing there has been no general acceptance of these terms, they may yet form the basis of a general classification scheme. The use of the term "surface" technique is also useful to describe the study of surface layers of the order of a hundred Ångstroms or so.

8.2   BASIS OF METHODS FOR THE STUDY OF CHEMICAL BONDING

Figure 8-1 illustrates the basis of the various methods for the study of chemical bonding. The atom is represented by several atomic levels including the K and L levels, plus the valence band. The valence band is shown to contain occupied and unoccupied levels, the demarkation between these being the Fermi level. An electron may be excited into the unoccupied levels (as in X-ray absorption) but to escape completely from the influence of its parent atom, it must also overcome the work function of the surface of the excited specimen. Case (a) shows normal photoelectron emission where a K electron has been ejected beyond the Fermi level, by a primary photon or electron of energy hv. The kinetic energy E_{kin} of the emitted photo-electron is given by

$$E_{kin} = hv - \phi_K \qquad (8\text{-}1)$$

where ϕ_K is the binding energy of the K shell. Strictly speaking the energy of E_{kin} given in equation (8-1) should be reduced by an amount of energy equivalent to the work function; however, since this correction is small, it can be ignored for the sake of simplicity.

Note that only *one* photoelectron occurs for each combination of hv and ϕ.

Case (b) illustrates the regaining of the ground state of the atom by transference of electrons from higher levels. This is the case dealt with in detail in

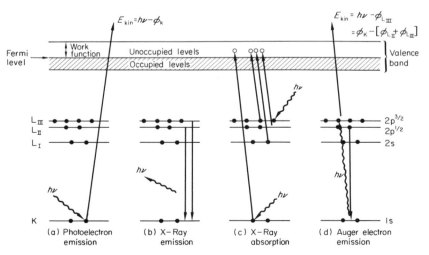

Fig. 8-1   Methods of determining binding energies (after Servier "Low Energy Electron Spectrometry")

Section 2-4 (pp. 16–18) and it is the phenomenon that leads to the production of characteristic X-ray quanta. The number of X-ray emission lines is determined by the selection rules already discussed. For the K series, the number of lines excited will be $2N$ where N is the total number of principal shells (*not* subshells) involved, which contain p electrons.

Case (c) illustrates the principle of photoelectric absorption. Here the impinging photons promote the orbital electrons to unoccupied levels. In doing so, the energy of the impinging photon is reduced by the amount $(\phi_x - \phi_y)$ where ϕ_x represents the binding energy of the level *from* which the electron was promoted and ϕ_y represents the binding energy in the unoccupied level *to* which the electron was promoted. The total photoelectric absorption τ total is given by

$$\tau_{\text{total}} = \sum [\tau_{xn}] \tag{8-2}$$

where τ_{xn} represents the absorption due to promotion of an electron from a particular subshell (see also equation (3-2)). Since the distribution of the excited electrons in the unoccupied levels will itself depend upon which levels are unoccupied, the absorption spectrum of a specimen will contain detail (fine structure) which reflects the availability of certain unoccupied levels.

Case (d) illustrates the de-excitation of the atom by the Auger process. In this instance, the X-ray photon, which arises following the transfer of an L shell electron to the K level does not escape, but in turn, ejects an electron from one of the L levels. There will be several combinations of L levels giving this effect, in fact in the simplest case, six such possibilities exist. Just one of these is illustrated, this being the $KL_{II}L_{III}$ transition. The energy of the X-ray photon is given by hv and when this ejects an electron from the L_{III} level, the kinetic energy given to this electron will be

$$E_{\text{kin}} = hv - \phi_{L_{III}}$$

where $\phi_{L_{III}}$ is the binding energy in the L_{III} level. Since hv is also equal to in this case, $\phi_K - \phi_{L_{II}}$ it follows that

$$E_{\text{kin}} = \phi_K - [\phi_{L_{II}} + \phi_{L_{III}}] \tag{8-3}$$

An important point to note is that the kinetic energy of the Auger electron is independent of the energy of the exciting radiation, although the latter must of course be great enough to excite an electron from the K level. As in case (a), the kinetic energy of the emitted Auger electron should be corrected for the work function.

8.3 INSTRUMENTATION FOR THE STUDY OF CHEMICAL BONDING

Figure 8-2 shows schematic diagrams of the types of equipment required for the study of chemical bonding. The equipment required for X-ray emission[5]

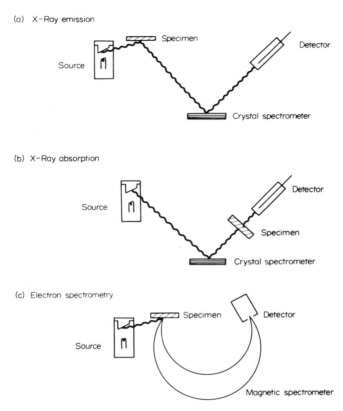

(a) X-Ray emission

Specimen

Detector

Source

Crystal spectrometer

(b) X-Ray absorption

Detector

Source

Specimen

Crystal spectrometer

(c) Electron spectrometry

Specimen   Detector

Source

Magnetic spectrometer

Fig. 8-2   Schematic diagrams showing the various methods which have been used for the study of chemical bonding

is almost identical to that required for conventional X-ray spectrometry, with one major exception. Since the X-ray emission method, as applied to bonding studies, is really only applicable to transitions in the outer orbitals and hence only to long wavelengths, it is commonplace to employ a special long wavelength X-ray source, typically an aluminium or copper target tube which gives in turn a high yield of aluminium K (~ 8 Å) or copper L (~ 13 Å) radiation. Also, the spectrometer will generally be required to measure wavelengths in the 10–100 Å, as opposed to the more usual 0.2–20 Å region covered in analytical X-ray spectrometry. The instrumentation will thus generally reflect this long wavelength requirement and will typically contain pseudo-crystals or gratings as dispersing agents.

The instrumentation required for X-ray absorption measurements is in its simplest form a powder diffractometer,[6] equipped with a single crystal in the usual specimen position and an absorption cell at the receiving slit. It is important to use an X-ray tube anode material which is significantly dissimilar in atomic number to that of the specimen. This is because X-ray tubes designed for powder diffractometry generally give a high characteristic wavelength output and this is a potential source of interference in the absorption measurement.

The basic instrumentation for all three forms of electron spectrometry is similar and differs only in the source of exciting radiation. In order to bring E_{kin} in equation (8-1) within reasonable limits, the source energy $h\nu$ is chosen such that it is of the order of several hundred electron volts in excess of ϕ. A typical PES source, therefore, would be an aluminium or magnesium target X-ray tube, possibly fitted with some form of crystal monochromator to reduce the source line-width. For u.v. source spectrometry, the source is generally a helium lamp giving resonance lines at 21.22 (He) and 40.81(He^+) eV, these lines in turn allowing the regions 0–20 eV and 20–40 eV to be excited. Although for Auger work many types of excitation source have been employed, including radioisotopes, X-ray photons, electrons and protons; the electron source is probably the most convenient. This has the main advantages of almost limitless controllability in terms of maximum energy, along with a greater ionisation cross-section than photoelectric excitation. Ultraviolet source spectrometry is generally carried out on specimens in the gaseous phase. Electron source Auger and PES are generally applied to specimens in the solid phase.

The spectrometer used in electron spectrometry is typically a magnetic or electrostatic system (retarding field systems have also been employed but are less common). Both types make use of the fact that moving electrons will be deflected from their trajectories by varying amounts, depending upon their kinetic energies. The magnetic system presents data in terms of number of electrons per unit momentum value versus momentum, and the electrostatic type in terms of unity energy value versus energy. The electrostatic type is generally commercially available and a typical ESCA system will allow the measurement of all forms of electron spectrometry. The spectrometer will generally cover the range zero to several thousand electron volts with a width at half maximum of about one electron volt. Facilities are generally available for the measurement of solids, liquids or gases. The detector is usually a Channeltron[8] (channel multiplier) or a Cu + Be photomultiplier, and data acquisition facilities of modern instrumentation[9] invariably include a small digital computer with some form of video output.

8.4   USE OF X-RAY EMISSION WAVELENGTHS

In Section 2.5 (p. 18) it was shown that the energy of an X-ray photon is proportional to the difference in the binding energies of the two levels involved in the appropriate transition. For instance, the Cu $K\alpha_1$ line is a $2p^{3/2} \rightarrow 1s$ transition (i.e., $L_{III} \rightarrow K$) and since the binding energies of these two levels are 9.984 keV and 0.936 keV respectively, the energy of the Cu $K\alpha_1$ photon will equal (9.984 − 0.936) keV = 8.048 keV. Since we are able to accurately measure the wavelength (and hence the energy) of the excited X-ray photons we are able, at least in principle, to study differences in binding level energies. Further, since the binding levels of the inner shells are to all intents and purposes constant, at least for all but the very low atomic number elements, this allows in turn the estimation of outer level binding energies.[10,11] As these outer levels are invariably involved in bonding, it should be theoretically possible to select a suitable X-ray wavelength and to use measured changes in this wavelength to study chemical bonding. In practice, however, this would require the measurement of a wavelength arising from an extreme orbital, for example, the $K\beta_2$ ($3p^{1/2} \rightarrow 1s$) line from copper and these wavelengths are generally poorly resolved from the other $K\beta$'s and also of low intensity. Further, the angular dispersion of the conventional X-ray spectrometer is relatively poor. The energy difference (dE) per angular increment (dθ) can be simply derived thus:

$$\lambda = \frac{2d}{n} \sin \theta = \frac{12{,}395}{E} \,(\text{eV})$$

$$\frac{\mathrm{d}E}{\mathrm{d}\theta} = \frac{12{,}395}{\lambda} \tan \theta$$

or

$$\mathrm{d}E = \frac{4.32}{\lambda} \tan \theta \,(\text{eV}) \qquad (8\text{-}4)$$

expressed for a dθ of 0.02°, which is roughly the angular precision of the conventional X-ray spectrometer. The smallest measurable angular increment would correspond to about 5 eV for 1 Å radiation and 0.5 eV for 10 Å radiation. This dispersion is barely sufficient for the measurement of the shorter wavelengths since a resolution of the order of better than 1 eV is generally required. The peak width at half height is also relatively poor since this will be determined by the natural mosaic spread of the crystal. Even in the very best case, the separation is never better than about 3 eV and this is about five times worse than can be *routinely* obtained with PES.

The shifts encountered in X-ray emission may be much smaller than those observed by PES, since the X-ray emission shifts represent only the *difference* between the shifts in levels and not the *total* shift of each level.

Figure 8-3 illustrates a consequence of the above and shows the spectra obtained from sodium thiosulphate with PES and X-ray emission. Sodium thiosulphate contains sulphur atoms in both 2^- and 6^+ oxidation states. In the ground state the electronic structure of sulphur is $KL3s^2 3p^4$. In the case of the central sulphur atom in the thiosulphate ion, the s and p orbitals hybridize to give six $s^2 p^4$ hybrid orbitals and a resulting sixfold covalent bond. The ligand sulphur atom is bonded by donation of two hybrid orbital electrons from the central sulphur into its two half-filled 3p orbitals. Hence the outer orbital electron density is different for the two sulphur atoms with a resulting difference

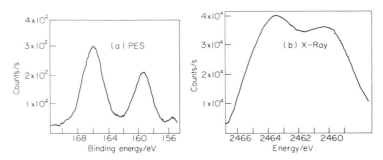

Fig. 8-3   Sulphur spectrum from sodium thiosulphate

in binding energy. The lower spectrum in Fig. 8-3 was obtained using a conventional X-ray spectrometer utilized under the very best conditions of angular dispersion and profile shape to record the $SK\beta$ line. Since this line is a $3p \rightarrow 1s$ transition, the transferred electron comes from one of the bonding orbitals described above.

It will be seen that the sulphur $K\beta$ line is broad and poorly defined and is in some ways similar to the $K\beta$ band spectra of aluminium previously discussed on p. 24. The upper spectrum was obtained with a PES spectrometer using magnesium $K\alpha$ radiation as the source.[12] The spectrum is of the 2p level and a shift of 6 eV is seen representing the two different sulphur atoms, since the central atom (S^{6+}) has a greater binding energy than the ligand atom (S^{2-}). The additional detail in the routinely obtained PES spectrum is clearly apparent and under conditions of high resolution it is also possible to observe line asymmetry due to the spin doublet splitting, i.e., $2p^{1/2}$ and $2p^{3/2}$.

The usefulness of the X-ray method using K series radiation increases for the very low atomic number elements where the energy gap between the K level and the bonding levels decreases. Needless to say, an equivalent situation will occur with somewhat higher atomic number elements where L series spectra are measured. As the energy gap between the K binding energy and the bonding levels decreases, i.e., with decrease in atomic number, a given energy gap represents a lower percentage of the average energy level. In other words, dE/E (the resolution of the spectrometer) increases with decrease of atomic number. This can also be seen by inspection of equation (8-1) where it is seen that dE is inversely proportional to λ. Evidence of this can be obtained by reference to Section 2 (p. 26) where mention was made of the oxygen K emission spectra; this is illustrated in Fig. 2.10. In this instance, the spectrometer has almost completely separated two maxima only 2 eV apart, since dE/E here is only 2/525 or 0.4%.

Study of the longer wavelength L emission spectrum from sulphur would thus be expected to give far more detail than the K spectrum and this is indeed so.[13] The sulphur L spectrum lies between 77 and 85 Å, or in terms of energy at around 150 eV, and sufficient detail is available to resolve the 3s, 3p and 3d bands.

The transition elements are typified by their partially filled 3d levels and so it is not surprising that much data has been published concerning the use of L emission spectra for study of these elements (see reference 2 for a detailed discussion of soft X-ray band spectra in metals). For instance, the 3d band in nickel has been found[14] to contain an unoccupied region of about 4 eV in width, involving two density of states maxima. This confirms previous measurements by paramagnetic susceptibility which indicates a similar situation in all transition metal from scandium through copper. A similar study[15] on the L emission bands from vanadium has been used to investigate non-stoichiometry in vanadium nitrides. Studies of this type are by no means restricted to the transition metal series and a great deal of work has been done on metals of relatively low atomic numbers, such as aluminium and magnesium. Since, however, the wavelengths of the L emission bands of these elements lie in the region of 200 Å, grating spectrometers are normally employed for recording spectra.

It will be apparent from what has gone previously, that, from specimens containing dissimilar atoms, the transferred electrons which come from bonding levels, may well arise from molecular orbitals, rather than hybridized atomic orbitals. Such a situation has already been discussed in Section 2 (p. 24) in the case of the $K\beta$ lines occurring in Al_2O_3 (see Fig. 2-8). Urch[16] has used this situation to study the π bonding between 3d and 2p orbitals in oxyanions. It is found from bond-length studies that 3d orbital participation must be present in the main-group elements in the second row of the periodic table. For instance, in a compound such as SF_6 which is assumed to contain six $d^2sp^2\sigma$ bonds, measurement of bond lengths shows that these lengths are too short to support the evidence of a pure σ bond. However, X-ray emission data can be used to show the presence of molecular orbital bands from which theoretical bond length calculations can be made which agree with experimental data.

8.5   ESCA METHODS

Some data have already been presented which illustrate the simplicity and high resolution of PES data. This is, however, not true of all ESCA methods, a case in point being Auger spectra. As an example, Fig. 8-4 shows typical spectra for oxygen obtained using X-ray, PES and Auger spectrometers. The X-ray K spectrum was obtained using a specimen of lithium carbonate, Li_2CO_3 and shows two broad maxima at about 524 eV and 526 eV. The two peaks probably correspond to the two types of oxygen, i.e., bonded to carbon or bonded to carbon and lithium, rather than to α_1 and α_2. The PES spectrum is the 1s of

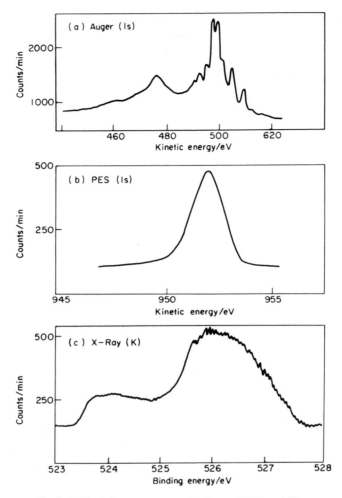

Fig. 8-4   Typical oxygen spectra by Auger, ESCA and X-ray

acetone CH_3COCH_3 and was obtained with Al $K\alpha$ radiation. The spectrum consists of one sharp, well-defined line. The Auger spectrum is of carbon sub-oxide C_3O_2 and in fact contains no less than eleven maxima.[17] The selection rules for Auger would, at first sight, only account for six lines since the K series transitions are expressed by KXpYg where X and Y are the electronic shells (in this case both X and Y are the L shell) and p and g are subshell indices. Figure 8-5 illustrates the transitions involved and Table 8-2 lists the X-ray, PES and the six Auger lines. It should be appreciated that although the six sets of transitions shown in Fig. 8-5 each appear to be dual transitions, they must in fact be considered instantaneous. It will be remembered that one of the

TABLE 8-2.
Types of radiation excited in oxygen

1. *X-Ray emission*
   The selection rules allow two lines:
   $K\alpha_1$ Energy $= (\phi_k - \phi_{L_{III}}) = (535 - 9\dagger) = 526 \text{ eV}\dagger$
   $K\alpha_2$ Energy $= (\phi_k - \phi_{L_{III}}) = (535 - 9\dagger) = 526 \text{ eV}\dagger$

2. *Photoelectron spectrometry*
   One line is observed for a fixed excitation energy, e.g., using an $AlK\alpha$ source ($hv = 1487 \text{ eV}$).
   $$E_{\text{kin}} = (hv - \phi_K) = (1487 - 535) = 952 \text{ eV}$$

3. *Auger spectrometry*
   Six lines given in Fig. 8-5 are as follows:
   KL_IL_I $E_{\text{kin}} = \phi_K - (\phi_{L_I} + \phi_{L_I}) = 532 - (29 + 29) = 474 \text{ eV}$
   KL_IL_{II} $E_{\text{kin}} = \phi_K - (\phi_{L_I} + \phi_{L_{II}}) = 532 - (29 + 14) = 489 \text{ eV}$
   KL_IL_{III} $E_{\text{kin}} = \phi_K - (\phi_{L_I} + \phi_{L_{III}}) = 532 - (29 + 11) = 492 \text{ eV}$
   $KL_{II}L_{II}$ $E_{\text{kin}} = \phi_K - (\phi_{L_{II}} + \phi_{L_{II}}) = 532 - (14 + 14) = 504 \text{ eV}$
   $KL_{II}L_{III}$ $E_{\text{kin}} = \phi_K - (\phi_{L_{II}} + \phi_{L_{III}}) = 532 - (14 + 11) = 507 \text{ eV}$
   $KL_{III}L_{III}$ $E_{\text{kin}} = \phi_K - (\phi_{L_{III}} + \phi_{L_{III}}) = 532 - (11 + 11) = 510 \text{ eV}$

† Approximate values.

selection rules is that $\Delta l = 1$ and if dual transitions did occur the first three sets of transitions would be "forbidden".

8.5.1 Different forms of orbital coupling

In order to understand the occurrence of the other Auger lines in the diagram it is necessary to have a more detailed knowledge of the types of atomic orbital interaction. In Section 2-6 (p. 20) the coupling of the l and s quantum numbers was considered, this giving the resultant vector sum j. This simple model is

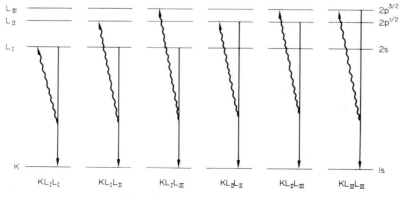

Fig. 8-5   KXpYq Auger transitions

sufficient to explain X-ray spectra in elementary terms but in more complex calculations involving binding electrons in atomic, hybridized or molecular orbitals it is necessary to be more specific in describing the types of coupling involved.[18] Two types of coupling are generally considered, these being "Russell-Saunders" coupling and "jj" coupling. In Russell-Saunders coupling, the orbits of individual electrons interact with each other more strongly than they interact with their spins. Thus the orbit vectors **l** can be considered to precess around their resultant **L**, and the spin vectors **s** precess around their resultant **S**. Thus

$$\sum l = \mathbf{L} \quad \text{and} \quad \sum s = \mathbf{S}$$

The resultant orbital moment **L** and spin moment **S**, precess in turn about each other forming a total moment **J**, or

$$\mathbf{J} = \mathbf{L} + \mathbf{S}$$

This represents the normal vector arrangement and is found to be a good approximation for lighter elements.

jj coupling represents the other extreme of the above, i.e., where orbit and spin vectors of an individual electron are coupled far more closely than with either moment of any other electrons. In this case, the individual orbit and spin vectors couple, i.e.,

$$j = l + s$$

and then the j vectors form a resultant **J** about which they precess, thus

$$\sum j = \mathbf{J}$$

jj coupling tends to predominate in heavier elements.

It was originally considered that Auger spectra arose mainly due to jj coupling since this would give the six K series lines already described. Not until the advent of higher resolution instruments was it appreciated that more lines exist and indeed using the concept of Russell-Saunders coupling nine lines would be predicted.[19] It is now considered that some form of intermediate coupling can best describe experimental results.[20]

8.5.2 Applications of ESCA methods

(i) Photoelectron spectrometry

The major application of PES is in inner core electron studies or PESIS. It has been found of great use both in analytical and applied chemistry and is equally applicable in organic and inorganic chemistry. At least five instruments are currently commercially available.[9] It has probably been most useful in structural chemistry where it has yielded a wealth of information on charge distributions, bond strengths and molecular arrangements. It is applicable, in principle at least, to any element in the periodic classification with the excep-

tions of hydrogen and helium. Generally, sensitivities lie in the region of 0.1 %
and the current state of the art allows quantitation to an accuracy of a few
percent.

(ii) Ultraviolet source spectrometry

The major application of u.v. source spectrometry is in outer core electron
studies PESOS.[21,22] Although its major fields of application are to be found
in the field of theoretical chemistry, it also has great uses in the analytical field
of gases and high vapour pressure liquids.

(iii) Auger spectrometry

Whereas PES and u.v. source spectrometry are relatively new techniques, the
principles of Auger spectrometry have been appreciated for nearly forty years.[23]
The analytical possibilities of the technique were, however, not realized until
1967 when Harris demonstrated greatly increased sensitivity.[24] Several good
reviews of the application are now available.[3,25,26] Auger spectroscopy can
be considered as genuine surface techniques since it is generally considered
that at electron energies of around 100 eV the first one or two atomic layers
are involved and at energies of 400 eV up to four atomic layers are concerned.
Since Auger electrons are most conveniently excited by electrons, the basic
instrumentation required for Auger work is similar to that required for LEED
(low energy electron diffraction). This combination is very powerful in that it
allows simultaneous compositional and structural data to be obtained.

The applications of Auger spectrometry are essentially all those involving
the study of surfaces and these include catalysis, diffusion studies in solids,
grain boundary segregation and migration of surface impurities.

8.6 ABSORPTION EDGE FINE STRUCTURE

Figure 8-1(c) illustrates that in the X-ray absorption process the electrons in
the absorber are excited to the unoccupied bonding levels beyond the Fermi
level. Since the position of the Fermi level will depend upon the number of
electrons already present in the bonding level, the energy gap between, for
example, the K level and the Fermi level will also vary with the number of
electrons present in the bonding level. Hence the position of absorption dis-
continuity for a given element may vary by a few tens of electron volts and further,
a certain amount of fine structure will occur at the short wavelength side of the
edge.

This fine structure will depend upon the availability of unoccupied quantum
states and the probability that the electron can undergo transitions to such
states. Absorption edge fine structure is found in crystalline and non-crystalline
solids, liquids and even polyatomic gases, and there have been many attempts
to provide an adequate theory to quantitatively explain experimental findings.[27]
most of the theories proposed are extensions of the original work by Kronig[28]

who, in 1931, predicted that in solids which crystallize with cubic symmetry, the absorption extremities are proportional to $[h^2 + k^2 + l^2]/a_0^2$, where h, k and l are the Miller indices and a_0 is the unit cell dimension. The premise here is that the scattering of the ejected photoelectron will depend upon the distribution of electrons in the crystal lattice. Kronig suggested that Bragg reflection of the ejected photoelectrons is responsible for the discontinuities in what would be a monotonic absorption curve for a free atom.

Strictly speaking, the Kronig theory deals only with so-called "long range order", which occurs up to several hundreds of electron volts from the absorption edge. This is essentially a phenomena which is restricted to crystalline

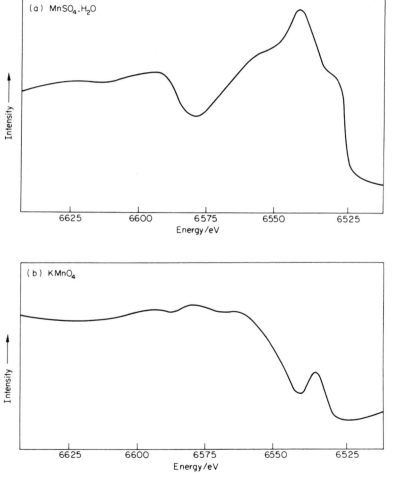

Fig. 8-6   Absorption edge spectra of manganese compounds

materials and does not explain the so-called "short range order" observed, for example, in gases. Short range order occurs over the range of a few tens of electron volts from the main absorption edge position, and is attributed to the transition of the photoelectron to the discrete energy states of an atom or molecule. This short range order is often called "Kossel structure".[29]

Although there has been a fair degree of success both in explaining and predicting the distribution of absorption edge fine structure in cubic materials,[30] there has been little success in the case of non-cubic materials. The calculations involve the calculation of the scattered wave in the x, y and z directions and only in cubic materials will these three be equivalent. Most of the practical

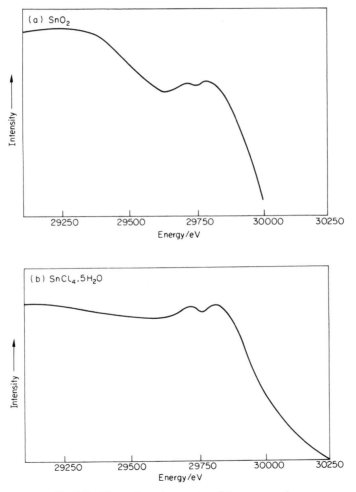

Fig. 8-7   Absorption edge spectra of tin compounds

applications of the use of absorption edge fine structure are, therefore, based on empirical comparisons between "unknown" specimens and reference standards containing elements in a "known" state of bonding.[31]

Figure 8-6 shows the absorption edge fine structure of $MnSO_4H_2O$ and $KMnO_4$. In the first of these compounds, the manganese is in the 2^+ state, and in the second in the 7^+ state. The average value of the K binding energy for manganese is 6539 eV and fine structure can be seen over the range 6525 to 6590 eV. In the 2^+ state, manganese has five unpaired electrons in the 3d level, but these are paired in the Mn configuration. This difference is probably the major cause for the shift in the discontinuities.

One of the major problems, even in the empirical interpretation of data, is the relatively poor resolution of the spectrometer. It will be seen from equation (8-4) that the resolution at the energy level of the manganese absorption edge ($\lambda = 1.89$ Å, $\theta = 44°$) is about 2.3 eV and in this instance it is sufficient to yield the required detail. For the shorter wavelengths, however, this degree of resolution is not attainable. For example, Fig. 8-7 illustrates the absorption edge spectra of SnO_2 and $SnCl_45H_2O$, obtained in the author's laboratory. Here the resolution is only about 10 eV and although gross differences are to be seen, these probably do not reflect the variations in electron density in the 5p level typified by these two compounds.

A further problem to be overcome in obtaining reproducible data is that of specimen preparation, since the measured intensity depends upon μ_x rather than on μ alone. It is unfortunate that the long wavelength region, where the resolution of the instrument is best, also happens to be that where specimen preparation is most difficult. In order to obtain good absorption spectra from elements such as manganese, a sample thickness of around 5–10 μm is required and in practice the thickness of such a specimen is very difficult to reproduce. Reasonable data is obtainable by soaking a filter paper with an aqueous solution of the salt in question, provided of course that it is, (a) water soluble and, (b) its absorption spectrum is unaffected by water of crystallization.

References

(1) Siegbahn, K., *et al.*, *ESCA Atomic Molecular and Solid State Structure Studied by means of Electron Spectroscopy*, Almquist and Wiksells, Uppsala, 1967; Siegbahn, K., *ESCA As Applied to Free Molecules*, Elsevier, New York, 1970.
(2) Fabian, D. J., (Ed.), *Soft X-ray Band Spectra*, Academic Press, London, 1968.
(3) Sevier, K. D., *Low Energy Electron Spectrometry*, Wiley, New York, 1971.
(4) Shirley, D. A., (Ed.), *Electron Spectroscopy*, North Holland, Amsterdam, 1972.
(5) Baun, W. L., *Appl. Spectry. Revs.* **1**, 379 (1968).
(6) Jenkins, R. and de Vries, J. L. *An Introduction to X-ray Powder Diffractometry*, Philips, Eindhoven, 1971.
(7) Carlson, T. A., in Shirley, D. A., (Ed.), *Electron Spectrometry*, North Holland, Amsterdam, 1972, p. 53.
(8) Sharber, J. R., Winningham, J. D. and Sheldon, W. R., *IEEE Trans. Nucl. Sci.* **NS-15**, 536 (1968).
(9) Karasek, F. W., *Research and Development* (Feb. 1972), p. 30.

(10) Urch, D. S., *Quart. Revs.*, **25**, 343 (1971).

(11) Faessler, A., *Angew. Chem. Intern. Ed. Engl.* **11**, 34 (1972).

(12) Siegbahn, K., *ESCA Atomic Molecular and Solid State Structure Studied by means of Electron Spectroscopy*, Almquist and Wiksells, Uppsala, 1967, p. 23.

(13) Fischer, D. W. and Baun, W. L., *Anal. Chem.* **37**, 902 (1965).

(14) Nemnonov, S. A., Volkov, V. F. and Suetin, V. S., *Fiz. Metal. i Metalloved.* **21**, 529 (1966) (translation VDC.439.292.548.73).

(15) Romand, M., Solomon, J. S. and Baun, W. L., *X-Ray Spectrom.* **1**, 147 (1972).

(16) Urch, D. S., *J. Chem. Soc. (A)*, 3026 (1969).

(17) Siegbahn, K., *ESCA Atomic Molecular and Solid State Structure Studied by means of Electron Spectroscopy*, Almquist and Wiksells, Uppsala, 1967, p. 23.

(18) Kuhn, H. G., *Atomic Spectra*, Longmans, London, 1961, p. 280.

(19) Siegbahn, K., *ESCA Atomic Molecular and Solid State Structure Studied by means of Electron Spectroscopy*, Almquist and Wiksells, Uppsala, 1967, appendix 4, p. 234.

(20) Asaad, W. N. and Burhop, E. H. S., *Proc. Phys. Soc. (London)* **71**, 369 (1958).

(21) Turner, D. W., Chem. Brit. 435 (1968).

(22) Riviere, J. C., Physics Bulletin, **20**, 85 (1969).

(23) Burhop, E. H. S., *The Auger Effect and Other Radiationless Transitions*, Cambridge University Press, Cambridge, 1952.

(24) Harris, L. A., *J. Appl. Phys.* **39**, 1419 (1968).

(25) Listengarten, M. A., *Bull. Acad. Sci. USSR, Phys. Ser.* **24**, 1050 (1960).

(26) Taylor, N. J., *J. Vacuum Sci. Technol.* **6**, 241 (1969).

(27) Azaroff, L. V., *Rev. Mod. Phys.* **35**, 1012 (1963).

(28) Kronig, R. de L., *Z. Physik* **70**, 317 (1931).

(29) Kossel, W., *Z. Physik* **1**, 119 (1920); **2**, 470 (1920).

(30) Sayers, D. E., Lytle, F. W. and Stern, E. A., *Advan. X-Ray Anal.* **13**, 248 (1969).

(31) Van Nostrand, R. A., *Advan. Catalysis* **12**, 149 (1960).

index

A

Absorption, 8
  curve, 39
  coefficient, 39
  critical depth, 41
  edge, 39–40, 41
  fine structure of, 153
  measurement of, 130
  primary, 41–43, 120
  secondary, 41–43, 120, 130
Alpha factors, 133
Analysis, *see* X-ray analysis
Analysing crystal, 47, 76–92
  angular dispersion, 47, 76, 81
  choice of, 88
  effect of temperature, 89
  ideally imperfect, 84
  mosaic spread, 82
Ångstrom unit, 8
Angular dispersion, *see* Analysing crystal
Angular reproducibility, goniometer, 88, *see also* X-Ray spectrometer
Atomic scattering factor, 50
Atomic structure factor, 39
Auger process, 17, 21, 144
Auger spectroscopy, 141, 153, *see also* Chemical bonding
Auger spectrum, 150
Autoionization, 17, 21, 27

B

Background, 41, 74, 114
Band spectra, 24, 25, 26, 54
Baseline restoration, 97
Beer–Lambert law, 131

Binding energy, 16, 19, 143
Bonding, *see* Chemical bonding
Bragg's law, 48
Burger–Dorgelo sum rule, 34

C

Calibration of spectrometer, 120
Chart recording, 100
Chemical bonding, study by X-ray methods 27, 141–156
  Auger spectroscopy, 149
  basis, 143
  comparison of methods, 140, 152
  ESCA, 149
  PES, 149
  X-ray emission, 146
Collimator, 41, 78
Correction procedures
  ALPHAS, 136
  CORSET, 128
  effective wavelength, 125
  empirical, 132, 138
  fundamental parameters, 123, 125
  graphical, 133
  in multi-element analysis, 136
  intensity/concentration algorithm, 128
  Lachance–Traill, 137
  Lucas–Tooth Price, 137
  regression analysis, 134
  secondary absorption, 130
  semi-empirical, 132
Continuum, 15
  intensity distribution of, 15
  maximum wavelength of, 15
  minimum wavelength of, 15, 41